Airy Functions and Applications to Physics

2nd Edition

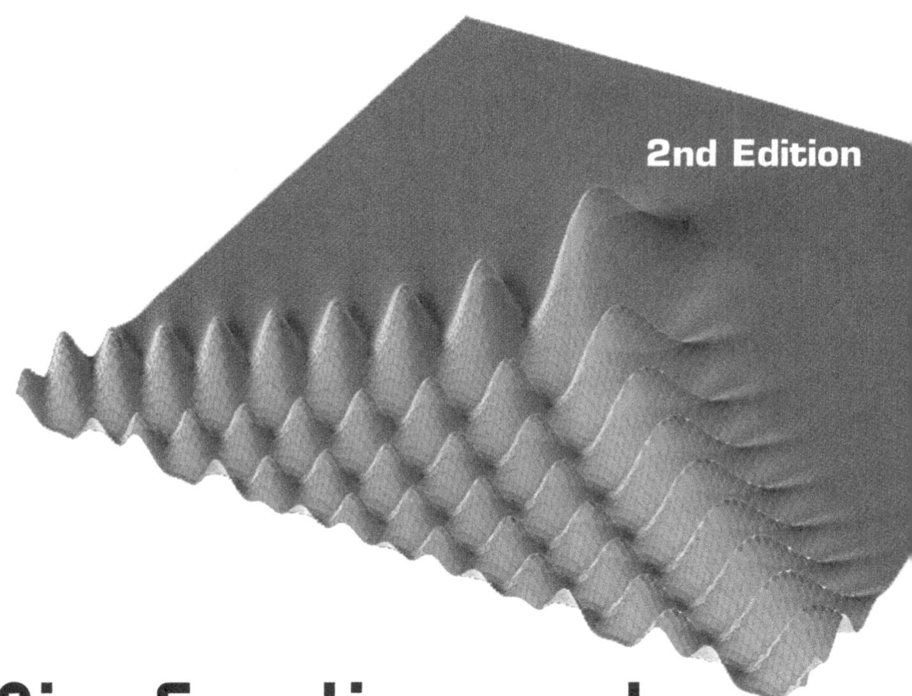

Airy Functions and Applications to Physics

Olivier Vallée
Université d'Orléans, France

Manuel Soares
DDEA des Yvelines, France

ICP
Imperial College Press

Published by

Imperial College Press
57 Shelton Street
Covent Garden
London WC2H 9HE

Distributed by

World Scientific Publishing Co. Pte. Ltd.
5 Toh Tuck Link, Singapore 596224
USA office: 27 Warren Street, Suite 401-402, Hackensack, NJ 07601
UK office: 57 Shelton Street, Covent Garden, London WC2H 9HE

British Library Cataloguing-in-Publication Data
A catalogue record for this book is available from the British Library.

AIRY FUNCTIONS AND APPLICATIONS TO PHYSICS (2nd Edition)

ISBN-13 978-1-84816-548-9
ISBN-10 1-84816-548-X

Printed in Singapore.

Preface

The use of special functions, and in particular of Airy functions, is rather common in physics. The reason may be found in the need to express a physical phenomenon in terms of an effective and comprehensive analytical form for the whole scientific community. However, for almost the last twenty years, many physical problems have been solved by computers, a trend which is now becoming the norm as the importance of computers continues to grow. Indeed as a last resort, the special functions employed in physics will have to be calculated numerically, even if the analytic formulation of physics is of prime importance.

Airy functions have periodically been the subject of many review articles, most of which focus on tabulations for the numerical calculation of these functions, as this is particularly difficult. The most well-known publications in this field are the tables by J. C. P. Miller, dating from 1946, and the chapter in the *Handbook of Mathematical Functions* by Abramowitz and Stegun, first published in 1954. Since that time, no noteworthy compilation of Airy functions has been published[1] in particular about the calculus involved in these functions. For example, in the latest editions of the tables by Gradshteyn and Ryzhik, they are hardly mentioned. Similarly, many results that have accrued over time in the scientific literature, remain extremely scattered and fragmentary.

Although Airy functions are used in many fields of physics, the analytical outcomes that have been obtained are not (or only poorly) transmitted among the various fields of research, which basically remain isolated. Moreover the tables by Abramowitz and Stegun are still the only common

[1] Note however that recently an equivalent compilation has become available on the web: the NIST Digital Library of Mathematical Functions – Chap. 9 Airy & Related Functions by F. W. J. Olver. Internet address: http://dlmf.nist.gov/

reference for all the authors using these functions. Thus many of the results have been rediscovered, and sometimes extremely old findings are the subject of publications, which represent a wasted effort for researchers.

The aim of this work is to make a rather exhaustive compilation of current knowledge on the analytical properties of Airy functions. In particular, the calculus involved in the Airy functions is carefully developed. This is, in fact, one of the major objectives of this book. While we are aware of repeating previous compilations to a large extent, this seemed necessary to ensure coherence. This book is addressed mainly to physicists (from advanced undergraduate students to researchers). We make no claim about the rigour of the mathematical demonstrations, as the reader will see.[2] The main aim is the result, or the fastest way to reach it. Finally, in the second part of this work, the reader will find some applications to various fields of physics. These examples are not exhaustive; they are only given to succinctly illustrate the use of Airy functions in classical or in quantum physics. For instance, we point out to the physicist in fluid mechanics that he can find what he is looking for in works on molecular physics or surface physics, and *vice versa*.

O. Vallée *&* M. Soares, Fall 2003

Preface to the second edition. In this second edition we have corrected various misprints of the first edition. Moreover we have added, here and there, new material such as catastrophe theory for the generalisation of the Airy function, additional results concerning the Airy transform and applications for instance to the Airy kernel, *etc.* It is our hope that this new edition will keep the book up to date in this still useful field.

We would like to thank Pr. Vladimir Varlamov and Pr. Sir Michael Berry for pointing out several omissions and flaws in the first edition.

O. Vallée *&* M. Soares, January 2010

[2]As a matter of fact, the Airy function can be considered as a distribution (generalised function) whose Fourier transform is an imaginary exponential. Also most of the integrals evoked in this work should be evaluated with the help of a convergence factor.

Contents

Preface v

1. A Historical Introduction: Sir George Biddell Airy 1

2. Definitions and Properties 5

2.1 Homogeneous Airy functions 5
2.1.1 The Airy equation 5
2.1.2 Elementary properties 7
2.1.3 Integral representations 9
2.1.4 Ascending and asymptotic series 11
2.2 Properties of Airy functions 16
2.2.1 Zeros of Airy functions 16
2.2.2 The spectral zeta function 17
2.2.3 Inequalities 20
2.2.4 Connection with Bessel functions 20
2.2.5 Modulus and phase of Airy functions 22
2.3 Inhomogeneous Airy functions 25
2.3.1 Definitions 25
2.3.2 Properties of inhomogeneous Airy functions . . . 27
2.3.3 Ascending series and asymptotic expansion 28
2.3.4 Zeros of the Scorer functions 29
2.4 Squares and products of Airy functions 30
2.4.1 Differential equation and integral representation . 30
2.4.2 A remarkable identity 32
2.4.3 The product $Ai(x)Ai(-x)$: Airy wavelets 32

3. Primitives and Integrals of Airy Functions 37

3.1 Primitives containing one Airy function 37
3.1.1 In terms of Airy functions 37
3.1.2 Ascending series 38
3.1.3 Asymptotic expansions 38
3.1.4 Primitives of Scorer functions 39
3.1.5 Repeated primitives 40
3.2 Product of Airy functions 40
3.2.1 The method of Albright 41
3.2.2 Some primitives 42
3.3 Other primitives 47
3.4 Miscellaneous 49
3.5 Elementary integrals 50
3.5.1 Particular integrals 50
3.5.2 Integrals containing a single Airy function 50
3.5.3 Integrals of products of two Airy functions 55
3.6 Other integrals 59
3.6.1 Integrals involving the Volterra μ-function 59
3.6.2 Canonisation of cubic forms 62
3.6.3 Integrals with three Airy functions 63
3.6.4 Integrals with four Airy functions 65
3.6.5 Double integrals 66

4. Transformations of Airy Functions 69

4.1 Causal properties of Airy functions 69
4.1.1 Causal relations 69
4.1.2 Green's function of the Airy equation 70
4.1.3 Fractional derivatives of Airy functions 72
4.2 The Airy transform 73
4.2.1 Definitions and elementary properties 73
4.2.2 Some examples 76
4.2.3 Airy polynomials 81
4.2.4 A particular case: correlation Airy transform . . . 83
4.3 Other kinds of transformations 94
4.3.1 Laplace transform of Airy functions 94
4.3.2 Mellin transform of Airy functions 95
4.3.3 Fourier transform of Airy functions 96
4.3.4 Hankel transform and the Airy kernel 97
4.4 Expansion into Fourier–Airy series 98

5. The Uniform Approximation 101

5.1 Oscillating integrals 101
5.1.1 The method of stationary phase 101
5.1.2 The uniform approximation of oscillating integrals 103
5.1.3 The Airy uniform approximation 104
5.2 Differential equations of the second order 104
5.2.1 The JWKB method 104
5.2.2 The Langer generalisation 106
5.3 Inhomogeneous differential equations 108

6. Generalisation of Airy Functions 111

6.1 Generalisation of the Airy integral 111
6.1.1 The generalisation of Watson 111
6.1.2 Oscillating integrals and catastrophes 114
6.2 Third order differential equations 118
6.2.1 The linear third order differential equation 118
6.2.2 Asymptotic solutions 119
6.2.3 The comparison equation 120
6.3 A differential equation of the fourth order 124

7. Applications to Classical Physics 127

7.1 Optics and electromagnetism 127
7.2 Fluid mechanics 130
7.2.1 The Tricomi equation 130
7.2.2 The Orr–Sommerfeld equation 132
7.3 Elasticity 135
7.4 The heat equation 137
7.5 Nonlinear physics 139
7.5.1 Korteweg–de Vries equation 139
7.5.2 The second Painlevé equation 143

8. Applications to Quantum Physics 147

8.1 The Schrödinger equation 147
8.1.1 Particle in a uniform field 147
8.1.2 The $|x|$ potential 151
8.1.3 Uniform approximation of the Schrödinger equation 154
8.2 Evaluation of the Franck–Condon factors 162

8.2.1 The Franck–Condon principle 163
8.2.2 The JWKB approximation 163
8.2.3 The uniform approximation 166
8.3 The semiclassical Wigner distribution 170
8.3.1 The Weyl–Wigner formalism 172
8.3.2 The one-dimensional Wigner distribution 173
8.3.3 The two-dimensional Wigner distribution 175
8.3.4 Configuration of the Wigner distribution 178
8.4 Airy transform of the Schrödinger equation 181

Appendix A Numerical Computation of the Airy Functions 185
A.1 Homogeneous functions 185
A.2 Inhomogeneous functions 187

Bibliography 191

Index 201

Chapter 1

A Historical Introduction: Sir George Biddell Airy

George Biddell Airy was born 27 July 1801 at Alnwick in Northumberland (North of England). His family was rather modest, but thanks to the generosity of his uncle Arthur Biddell, he went to study at Trinity College, University of Cambridge. Although a sizar,[1] he was a brilliant student and finally graduated in 1823 as a senior wrangler. Three years later he was appointed to the celebrated Lucasian chair of mathematics. However, his salary as Lucasian professor was too small to marry Richarda, the young lady's father objected, so he applied for a new position. In 1828, Airy obtained the Plumian chair, becoming professor of Astronomy and director of the new observatory at Cambridge. His early work at this time concerned the mass of Jupiter and the irregular motions of the Earth and Venus.

In 1834, Airy started his first mathematical studies on the diffraction phenomenon and optics. Due to diffraction, the image of a point through a telescope is actually a spot surrounded by rings of smaller intensity. This spot is now called the *Airy spot*; the associated Airy function, however, has nothing to do with the purpose of this book.

In June 1835, Airy became the 7th Royal Astronomer and director of the Greenwich observatory, succeeding John Pond. Under his administration, modern equipment was installed, leading the observatory to worldwide fame assisted by the quality of its published data. Airy also introduced the study of sun spots and of the Earth's magnetism, and built new apparatus for the observation of the Moon, and for cataloguing the stars. The question of absolute time was also a major challenge: Airy defined the *Airy Transit Circle*, which in 1884 became the Greenwich Mean Time. However the renown of Airy is also due to the *Neptune affair*. During the decade 1830–

[1]Meaning that he paid a reduced fee in exchange for working as a servant to richer students.

40, astronomers were interested in the perturbations of Uranus, which had been discovered in 1781. In France, François Arago suggested to Urbain Le Verrier that he should seek for a new planet that might have caused the perturbations of Uranus. In England, the young John Adams was doing the same calculations with a slight advance. Airy however was dubious about the outcome of his work. Adams tried twice to meet Airy in 1845 but was unsuccessful: the first time Airy was away, the second time Airy was having dinner and did not wish to be disturbed. Finally, Airy entrusted the astronomer James Challis with the observation of the new planet from the calculations of Adams. Unfortunately, Challis failed in his task. At the same time, Le Verrier asked the German astronomer Johann Galle in Berlin to locate the planet from his data: the new planet was discovered on 20 September 1846. A controversy then started between Airy and Arago, between France and England, and also against Airy himself. The dispute became more acrimonious concerning the name of the planet itself, Airy wanting to name the new planet Oceanus. The name Neptune was finally given. The story goes that, in the end, Adams and Le Verrier became good friends.

In 1854 Airy attempted to determine the mean density of the Earth by comparing the gravity forces on a single pendulum at the top and the bottom of a pit. The experiment was carried out near South Shields in a mine 1250 feet in depth. Taking into account the elliptical form and the rotation of the Earth, Airy deduced a density of 6.56, which is not so far — considering the epoch — from the currently accepted density 5.42.

Airy was knighted in 1872, and so became Sir George Biddell.[2] At this time, Airy started a lunar theory. The results were published in 1886, but in 1890 he found an error in his calculations. The author was then 89 years old and was unwilling to revise his calculations. Late in 1881, Sir George retired from his position as astronomer at Greenwich. He died January 2, 1892.

The autobiography of Sir George, edited by his son Wilfred, was published in 1896 [W. Airy (1896)]. The name of Airy is associated with many phenomena such as the Airy spiral (an optical phenomenon visible in quartz crystals), the Airy spot in diffraction phenomena or the Airy stress function which he introduced in his work on elasticity, which is different again from the Airy functions that we shall discuss in this book. Among the most well-known books he wrote, we may mention *Mathematical tracts on physical astronomy* (1826) and *Popular astronomy* (1849) [W. Airy (1896)].

[2] After having declined the offer on three occasions, objecting to the fees.

Fig. 1.1 Sir George Biddell Airy (after the Daily Graphic, January 6, 1892).

Airy was particularly involved in optics: for instance, he made special glasses to correct his own astigmatism. For the same reason, he was also interested in the calculation of light intensity in the neighbourhood of a caustic [Airy (1838), (1849)]. For this purpose, he introduced the function defined by the integral

$$W(m) = \int_0^\infty \cos\left[\frac{\pi}{2}\left(\omega^3 - m\omega\right)\right] \mathrm{d}\omega,$$

which is now called the Airy function. This is the object of the present book. W is the solution to the differential equation

$$W'' = -\frac{\pi^2}{12} mW.$$

The numerical calculation of Airy functions is somewhat tricky, even today! However in 1838, Airy gave a table of the values of W for m varying from -4.0 to $+4.0$. Thence in 1849, he published a second table for m varying from -5.6 to $+5.6$, for which he employed the ascending series. The problem is that this series is slowly convergent as m increases. A few years later, Stokes (1851, 1858) introduced the asymptotic series of $W(m)$, of its derivative and of the zeros. Practically no research was undertaken on the Airy function until the work by Nicholson (1909), Brillouin (1916) and Kramers (1926) who contributed significantly to our knowledge of this function.

In 1928 Jeffreys introduced the notation used nowadays

$$Ai(x) = \frac{1}{\pi} \int_0^{\infty} \cos\left(\frac{t^3}{3} + xt\right) \mathrm{d}t,$$

which is the solution of the homogeneous differential equation, called the Airy equation

$$y'' = xy.$$

Clearly, this equation may be considered as an approximation of the differential equation of the second order

$$y'' + F(x)y = 0,$$

where F is any function of x. If $F(x)$ is expanded in the neighbourhood of a point $x = x_0$, we have to the first order $(F'(x_0) \neq 0)$

$$y'' + [F(x_0) + (x - x_0)F'(x_0)]\, y = 0.$$

Then with a change of variable, we find the Airy equation. This method is particularly useful in the neighbourhood of a zero of $F(x)$. The point x_0 defined by the relation $F(x_0) = 0$ is called a transition point by mathematicians and a turning point by physicists. Turning points are involved in the asymptotic solutions to linear differential equations of the second order [Jeffreys (1942)], such as the stationary Schrödinger equation.

Finally we can note that Airy functions are Bessel functions (or linear combinations of these functions) of order 1/3. The relation between the Airy equation and the Bessel equation is performed with the change of variable $\xi = \frac{2}{3}x^{3/2}$, leading Jeffreys (1942) to say: "*Bessel functions of order* 1/3 *seem to have no application except to provide an inconvenient way of expressing this function*!"

Chapter 2

Definitions and Properties

This chapter is devoted to general definitions and properties of Airy functions as they can be, at least partially, found in the chapter concerning these functions in the *Handbook of Mathematical Functions* by Abramowitz & Stegun (1965).

2.1 Homogeneous Airy functions

2.1.1 *The Airy equation*

We consider the following homogeneous second order differential equation, called the Airy equation

$$y'' - xy = 0. \tag{2.1}$$

This differential equation may be solved by the method of Laplace, i.e. in seeking a solution as an integral

$$y = \int_{\mathrm{C}} \mathrm{e}^{xz} v(z) \mathrm{d}z,$$

which is equivalent to solving the first order differential equation

$$v' + z^2 v = 0.$$

We thus obtain the solution to Eq. (2.1), except for a normalisation constant,

$$y = \int_{\mathrm{C}} \mathrm{e}^{xz - z^3/3} \mathrm{d}z.$$

The integration path C is chosen such that the function $v(z)$ must vanish at the boundaries. This is the reason why the extremities of the path must

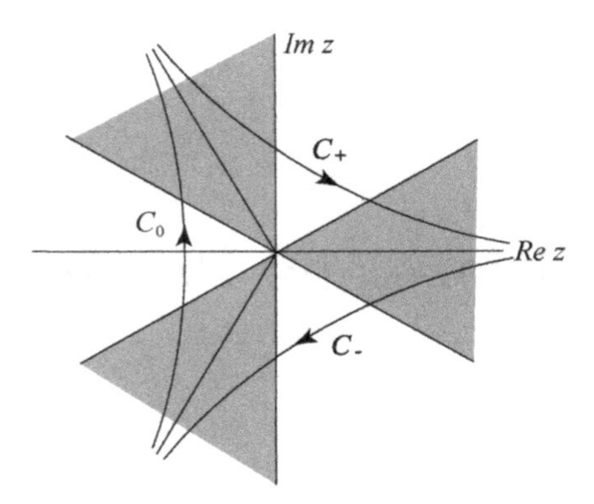

Fig. 2.1 Integration paths associated to the solution of the Airy equation (2.1).

tend to infinity in the regions of the complex plane z, where the real part of z^3 is positive (shaded regions of the complex plane).

From symmetry considerations, it is useful to work with the paths C_0, C_+ and C_-. Clearly the integration paths C_+ and C_- lead to solutions that tend to infinity when x goes to infinity. When we consider the path C_0 and the associated solution, we can deform this curve until it joins the imaginary axis. We now define the Airy function Ai by

$$Ai(x) = \frac{1}{2\pi \mathrm{i}} \int_{-\mathrm{i}\infty}^{+\mathrm{i}\infty} \mathrm{e}^{xz - z^3/3} \mathrm{d}z. \tag{2.2}$$

If $1, \mathrm{j}, \mathrm{j}^2$ are the cubic roots of unity (that is to say $\mathrm{j} = \mathrm{e}^{\mathrm{i}2\pi/3}$) the functions defined by the paths C_+ and C_- are respectively the functions $Ai(\mathrm{j}x)$ and $Ai(\mathrm{j}^2x)$. Combining these solutions, two by two linearly independents for they satisfy the same second order differential equation, we have the relation

$$Ai(x) + \mathrm{j}Ai(\mathrm{j}x) + \mathrm{j}^2 Ai(\mathrm{j}^2 x) = 0. \tag{2.3}$$

Now, in place of the functions $Ai(\mathrm{j}x)$ and $Ai(\mathrm{j}^2x)$, we define the function $Bi(x)$, linearly independent of $Ai(x)$, which has the interesting property of being real when x is real

$$Bi(x) = \mathrm{i}\mathrm{j}^2 Ai(\mathrm{j}^2 x) - \mathrm{i}\mathrm{j}Ai\left(\mathrm{j}x\right). \tag{2.4}$$

Similarly to $Ai(x)$ (cf. formula (2.3)), we have the relation

$$Bi(x) + \mathrm{j}Bi(\mathrm{j}x) + \mathrm{j}^2 Bi(\mathrm{j}^2 x) = 0. \tag{2.5}$$

On Figs. (2.2) and (2.3), the plots of the functions $Ai(x)$, $Bi(x)$, and of their derivatives $Ai'(x)$ and $Bi'(x)$ are given.

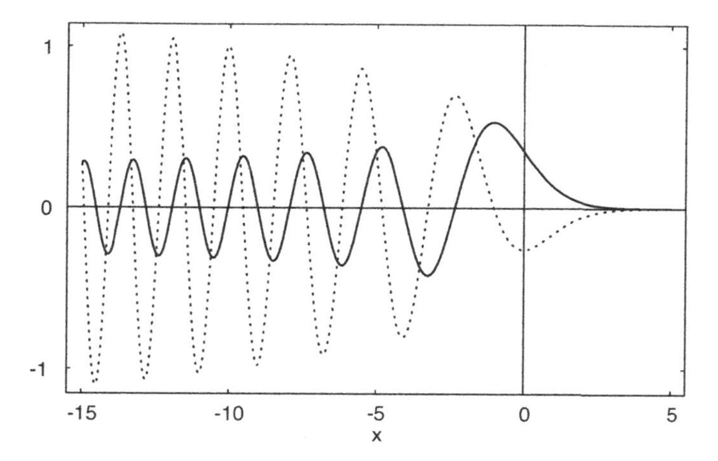

Fig. 2.2 Plot of the Airy function Ai (solid line) and its derivative (dotted line).

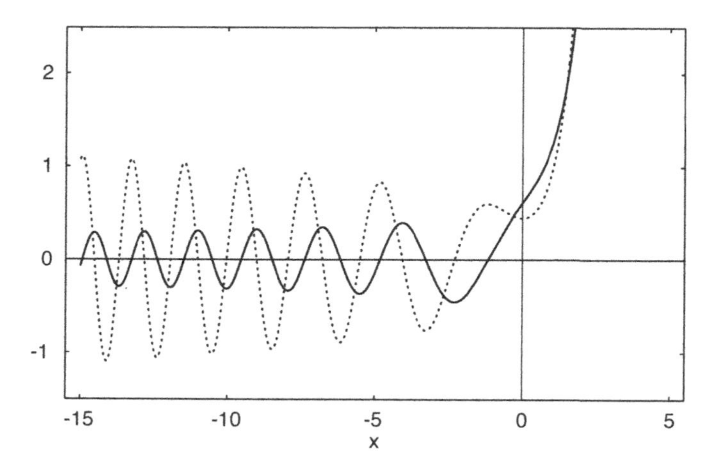

Fig. 2.3 Plot of the Airy function Bi (solid line) and its derivative (dotted line).

2.1.2 *Elementary properties*

2.1.2.1 *Wronskians of homogeneous Airy functions*

The Wronskian $W\{f,g\}$ of two functions $f(x)$ and $g(x)$ is defined by

$$W\{f,g\} = f(x)\frac{\mathrm{d}g(x)}{\mathrm{d}x} - \frac{\mathrm{d}f(x)}{\mathrm{d}x}g(x).$$

For the Airy functions Ai and Bi, we have the following Wronskians [Abramowitz & Stegun (1965)]

- $$W\{Ai(x),\ Bi(x)\} = \frac{1}{\pi} \tag{2.6}$$
- $$W\left\{Ai(x),\ Ai\left(x\mathrm{e}^{\mathrm{i}2\pi/3}\right)\right\} = \frac{\mathrm{e}^{-\mathrm{i}\pi/6}}{2\pi} \tag{2.7}$$
- $$W\left\{Ai(x),\ Ai\left(x\mathrm{e}^{-\mathrm{i}2\pi/3}\right)\right\} = \frac{\mathrm{e}^{\mathrm{i}\pi/6}}{2\pi} \tag{2.8}$$
- $$W\left\{Ai\left(x\mathrm{e}^{\mathrm{i}2\pi/3}\right),\ Ai\left(x\mathrm{e}^{-\mathrm{i}2\pi/3}\right)\right\} = \frac{\mathrm{i}}{2\pi}. \tag{2.9}$$

2.1.2.2 *Particular values of Airy functions*

The values at the origin of homogeneous Airy functions are

$$Ai(0) = \frac{Bi(0)}{\sqrt{3}} = \frac{1}{3^{2/3}\Gamma\left(\frac{2}{3}\right)} = 0.355\,028\,053\,887\,817\,239 \tag{2.10}$$

$$-Ai'(0) = \frac{Bi'(0)}{\sqrt{3}} = \frac{1}{3^{1/3}\Gamma\left(\frac{1}{3}\right)} = 0.258\,819\,403\,792\,806\,798 \tag{2.11}$$

and therefore

$$Ai(0)Ai'(0) = \frac{-1}{2\pi\sqrt{3}}. \tag{2.12}$$

More generally, we have for the higher derivatives [Crandal (1996)]

$$Ai^{(n)}(0) = (-1)^n c_n \sin(\pi(n+1)/3), \tag{2.13}$$

and

$$Bi^{(n)}(0) = c_n[1 + \sin(\pi(4n+1)/6)], \tag{2.14}$$

where the coefficient c_n is

$$c_n = \frac{1}{\pi} 3^{(n-2)/3}\Gamma\left(\frac{n+1}{3}\right).$$

2.1.2.3 *Relations between Airy functions*

The following relations are deduced from formulae (2.3), (2.4) and (2.5) [Miller (1946); Abramowitz & Stegun (1965)]

- $$Ai\left(x\mathrm{e}^{\pm\mathrm{i}2\pi/3}\right) = \frac{\mathrm{e}^{\pm\mathrm{i}\pi/3}}{2}\left[Ai(x) \mp \mathrm{i}Bi(x)\right] \tag{2.15}$$
- $$Ai'\left(x\mathrm{e}^{\pm\mathrm{i}2\pi/3}\right) = \frac{\mathrm{e}^{\mp\mathrm{i}\pi/3}}{2}\left[Ai'(x) \mp \mathrm{i}Bi'(x)\right] \tag{2.16}$$
- $$Bi(x) = \mathrm{e}^{\mathrm{i}\pi/6} Ai\left(x\mathrm{e}^{\mathrm{i}2\pi/3}\right) + \mathrm{e}^{-\mathrm{i}\pi/6} Ai\left(x\mathrm{e}^{-\mathrm{i}2\pi/3}\right) \tag{2.17}$$
- $$Bi'(x) = \mathrm{e}^{\mathrm{i}5\pi/6} Ai'\left(x\mathrm{e}^{\mathrm{i}2\pi/3}\right) + \mathrm{e}^{-\mathrm{i}5\pi/6} Ai'\left(x\mathrm{e}^{-\mathrm{i}2\pi/3}\right). \tag{2.18}$$

Finally we have this interesting relation

$$\Re[Ai(jx)Bi(j^2x)] = Ai(x)Bi(x), \tag{2.19}$$

where $\Re$ means the real part.

2.1.3 *Integral representations*

An integral definition of $Ai(x)$ was given by formula (2.2). This function can also be defined by the following formulae [Abramowitz & Stegun (1965)]

- $$Ai(x) = \frac{1}{\pi}\int_0^{\infty} \cos\left(\frac{z^3}{3} + xz\right) \mathrm{d}z \tag{2.20}$$
- $$Ai(x) = \frac{1}{2\pi}\int_{-\infty}^{+\infty} \mathrm{e}^{\,\mathrm{i}(z^3/3+xz)}\mathrm{d}z \tag{2.21}$$
- $$Ai(x) = \frac{x^{1/2}}{2\pi}\int_{-\infty}^{+\infty} \mathrm{e}^{\,\mathrm{i}x^{3/2}(z^3/3+z)}\mathrm{d}z,\ x > 0 \tag{2.22}$$
- $$Ai(x) = \frac{\mathrm{e}^{-\xi}}{2\pi}\int_{-\infty}^{+\infty} \mathrm{e}^{-xz^2+\mathrm{i}z^3/3}\mathrm{d}z,\ x > 0,\ \xi = \frac{2}{3}x^{3/2} \tag{2.23}$$
- $$Ai(x) = \frac{\mathrm{e}^{-\xi}}{\pi}\int_0^{\infty} \mathrm{e}^{-xz^2}\cos\left(\frac{z^3}{3}\right) \mathrm{d}z,\ x > 0,\ \xi = \frac{2}{3}x^{3/2}. \tag{2.24}$$

More generally, we have

$$Ai(a\,x) = \frac{1}{2\pi a}\int_{-\infty}^{+\infty} \exp\left[\mathrm{i}\left(\frac{u^3}{3\,a^3} + xu\right)\right]\mathrm{d}u. \tag{2.25}$$

Note also the useful formula

$$\int_{-\infty}^{+\infty} \exp\left[\mathrm{i}\left(\frac{t^3}{3} + a\,t^2 + bt\right)\right]\mathrm{d}t = 2\pi\,\mathrm{e}^{\mathrm{i}a(2a^2/3-b)}\,Ai(b - a^2). \tag{2.26}$$

Olver (1974) gives the expressions

$$Ai(x) = \frac{1}{\mathrm{i}\pi}\int_0^{\mathrm{i}\infty} \cosh\left(\frac{z^3}{3} - xz\right) \mathrm{d}z, \tag{2.27}$$

which for $x > 0$ gives

$$Ai(-x) = \frac{x^{1/2}}{\pi} \int_{-1}^{\infty} \cos\left[x^{3/2}\left(\frac{z^3}{3} + z^2 - \frac{2}{3}\right)\right] dz. \tag{2.28}$$

Copson (1963) gives the expression

$$Ai(x) = \frac{e^{-\xi}}{2\pi} \int_0^{\infty} e^{-x^{1/2} z} \cos\left(\frac{z^{3/2}}{3}\right) \frac{dz}{\sqrt{z}}, \quad x > 0, \ \xi = \frac{2}{3} x^{3/2}, \tag{2.29}$$

and Reid (1995) the following expression ($x > 0$)

$$\begin{aligned} Ai(x) &= \frac{\sqrt{3}}{2\pi} \int_0^{+\infty} e^{-\frac{x^3 t^3}{3} - \frac{1}{3t^3}} \frac{dt}{t^2} \\ &= \frac{\sqrt{3}}{2\pi} \int_0^{+\infty} e^{-\frac{t^3}{3} - \frac{x^3}{3t^3}} dt. \end{aligned} \tag{2.30}$$

For the function Bi, we have the integral representation

$$Bi(x) = \frac{1}{\pi} \int_0^{\infty} \left[e^{-z^3/3 + xz} + \sin\left(\frac{z^3}{3} + xz\right)\right] dz. \tag{2.31}$$

Other formulae [Gordon (1969); Schulten *et al.* (1979)] of great interest for the numerical computation of Airy functions are obtained by setting

$$\rho(x) = \frac{1}{\pi^{1/2}\, 2^{11/6}\, 3^{2/3}\, x^{2/3}} e^{-x} Ai\left[\left(\frac{3x}{2}\right)^{2/3}\right], \quad x > 0. \tag{2.32}$$

In fact, the Bessel function $K_\nu(x)$ verifies the relation [Gradshteyn & Ryzhik (1965)]

$$\int_0^{\infty} \frac{e^{-u} K_\nu(u)}{u + t} \frac{du}{\sqrt{u}} = \pi \frac{e^t K_\nu(t)}{\sqrt{t} \cos(\pi\nu)}, \quad \Re(\nu) < \frac{1}{2}, \ \arg(t) < \pi.$$

In particular, for $K_{1/3}(x) = \frac{\pi\sqrt{3}}{\left(\frac{3x}{2}\right)^{1/3}} Ai\left[\left(\frac{3x}{2}\right)^{2/3}\right]$ (cf. §2.2.4), we obtain

$$Ai(x) = \frac{e^{-\xi}}{2\pi^{1/2} x^{1/4}} \int_0^{+\infty} \frac{\rho(z)}{1 + \frac{z}{\xi}} dz, \tag{2.33}$$

with ρ defined as above and $\xi = \frac{2}{3}x^{3/2}$. In a similar fashion, we have for $x > 0$

$$Bi(x) = \frac{\mathrm{e}^{\xi}}{\pi^{1/2}x^{1/4}} \int_0^{+\infty} \frac{\rho(z)}{1-\frac{z}{\xi}} \mathrm{d}z \tag{2.34}$$

$$Ai(-x) = \frac{1}{\pi^{1/2}x^{1/4}} \int_0^{+\infty} \frac{\cos\left(\xi - \frac{\pi}{4}\right) + \frac{z}{\xi}\sin\left(\xi - \frac{\pi}{4}\right)}{1+\left(\frac{z}{\xi}\right)^2} \rho(z)\mathrm{d}z \tag{2.35}$$

$$Bi(-x) = \frac{1}{\pi^{1/2}x^{1/4}} \int_0^{+\infty} \frac{\frac{z}{\xi}\cos\left(\xi - \frac{\pi}{4}\right) - \sin\left(\xi - \frac{\pi}{4}\right)}{1+\left(\frac{z}{\xi}\right)^2} \rho(z)\mathrm{d}z. \tag{2.36}$$

It should be noted that the integral representations (2.20) and (2.21) are the most frequently used.

2.1.4 *Ascending and asymptotic series*

2.1.4.1 *Expansion of Ai near the origin*

The expansion of Ai near the origin $x = 0$ is [Copson (1967)]

$$Ai(x) = \frac{1}{\pi 3^{2/3}} \sum_{n=0}^{\infty} \frac{\Gamma\left(\frac{n+1}{3}\right)}{n!} \sin\left[\frac{2}{3}(n+1)\pi\right] \left(3^{1/3}x\right)^n. \tag{2.37}$$

2.1.4.2 *Ascending series of Ai and Bi*

The ascending series of $Ai(x)$ and $Bi(x)$ are defined [Miller (1946); Abramowitz & Stegun (1965)] by the following chain rule

$$Ai(x) = c_1 f(x) - c_2 g(x) \tag{2.38}$$

$$Bi(x) = \sqrt{3}\left[c_1 f(x) + c_2 g(x)\right], \tag{2.39}$$

with $c_1 = Ai(0)$ and $c_2 = -Ai'(0)$, and the series

$$f(x) = \sum_{k=0}^{\infty} 3^k \left(\frac{1}{3}\right)_k \frac{x^{3k}}{(3k)!} = 1 + \frac{1}{3!}x^3 + \frac{1.4}{6!}x^6 + \frac{1.4.7}{9!}x^9 + \ldots$$

$$g(x) = \sum_{k=0}^{\infty} 3^k \left(\frac{2}{3}\right)_k \frac{x^{3k+1}}{(3k+1)!} = x + \frac{2}{4!}x^4 + \frac{2.5}{7!}x^7 + \frac{2.5.8}{10!}x^{10} + \ldots$$

where the Pochhammer symbol $(a)_n$ is defined by

$$(a)_0 = 1, \; (a)_n = \frac{\Gamma(a+n)}{\Gamma(a)} = a(a+1)(a+2)\ldots(a+n-1). \tag{2.40}$$

The ascending series of the derivatives $Ai'(x)$ and $Bi'(x)$ are obtained from the differentiation of the series $f(x)$ and $g(x)$ term-by-term. We obtain therefore

$$Ai'(x) = c_1 f'(x) - c_2 g'(x) \tag{2.41}$$

$$Bi'(x) = \sqrt{3}\left[c_1 f'(x) + c_2 g'(x)\right], \tag{2.42}$$

and the series

$$f'(x) = \frac{x^2}{2} + \frac{1}{2.3}\frac{x^5}{5} + \frac{1}{2.3.5.6}\frac{x^8}{8} + \ldots$$

$$g'(x) = 1 + \frac{1}{1.3}\frac{x^3}{3} + \frac{1}{1.3.4.6}\frac{x^6}{6} + \frac{1}{1.3.4.6.7.9}\frac{x^9}{9} + \ldots.$$

2.1.4.3 *Asymptotic expansion of Ai and Bi*

The asymptotic expansions of Ai and Bi are calculated with the steepest descent method [Olver (1954); Chester *et al.* (1957)]. We will calculate the asymptotic expansion of $Ai(x)$ for $x > 0$. Definition (2.2) of Ai allows us to write

$$Ai(x) = \frac{1}{2\pi i}\int_{C_0} e^{t^3/3 - xt} dt,$$

where C_0 is the contour defined in §2.1.1. Setting

$$t = \sqrt{x} + iu, \; -\infty < u < \infty,$$

we obtain

$$\pi e^{\xi} Ai(x) = \int_0^{\infty} e^{-u^2\sqrt{x}} \cos\left(\frac{u^3}{3}\right) du$$

$$= \frac{1}{2x^{1/4}}\int_{-\infty}^{+\infty} e^{-v^2} \cos\left(\frac{v^3}{3x^{3/4}}\right) dv,$$

with $\xi = \frac{2}{3}x^{3/2}$. The *cosine* function may be replaced by its expansion:

$$\pi \mathrm{e}^{\xi} Ai(x) = \frac{1}{2x^{1/4}} \int_{-\infty}^{+\infty} \mathrm{e}^{-v^2} \left(1 - \frac{v^6}{2!3^2 x^{3/2}} + \frac{v^{12}}{4!3^4 x^3} - \dots\right) \mathrm{d}v.$$

Integrating term–by–term

$$\pi \mathrm{e}^{\xi} Ai(x) \approx \frac{\pi^{1/2}}{2x^{1/4}} \left(1 - \frac{3.5}{1!144 x^{3/2}} + \frac{5.7.9.11}{2!144^2 x^3} - \dots\right),$$

we obtain the formula given below (2.45). The other expansions are evaluated similarly. For $x \gg 1$ and $s \geq 1$, one defines

$$u_s = \frac{\Gamma(3s+1/2)}{54^s s! \Gamma(s+1/2)} = \frac{(2s+1)(2s+3)\dots(6s-1)}{216^s s!} \tag{2.43}$$

$$v_s = -\frac{6s+1}{6s-1} u_s, \tag{2.44}$$

and the other expansion, according to the notation of Olver (1954)

$$L(z) = \sum_{s=0}^{\infty} \frac{u_s}{z^s} = 1 + \frac{3.5}{1!216} \frac{1}{z} + \frac{5.7.9.11}{2!216^2} \frac{1}{z^2} + \frac{7.9.11.13.15.17}{3!216^3} \frac{1}{z^3} + \dots$$

$$M(z) = \sum_{s=0}^{\infty} \frac{v_s}{z^s} = 1 - \frac{3.7}{1!216} \frac{1}{z} - \frac{5.7.9.13}{2!216^2} \frac{1}{z^2} - \frac{7.9.11.13.15.19}{3!216^3} \frac{1}{z^3} - \dots$$

$$P(z) = \sum_{s=0}^{\infty} (-1)^s \frac{u_{2s}}{z^{2s}} = 1 - \frac{5.7.9.11}{2!216^2} \frac{1}{z^2} + \frac{9.11.13.15.17.19.21.23}{4!216^4} \frac{1}{z^4} - \dots$$

$$Q(z) = \sum_{s=0}^{\infty} (-1)^s \frac{u_{2s+1}}{z^{2s+1}} = \frac{3.5}{1!216} \frac{1}{z} - \frac{7.9.11.13.15.17}{3!216^3} \frac{1}{z^3} + \dots$$

$$R(z) = \sum_{s=0}^{\infty} (-1)^s \frac{v_{2s}}{z^{2s}} = 1 + \frac{5.7.9.13}{2!216^2} \frac{1}{z^2} - \frac{9.11.13.15.17.19.21.25}{4!216^4} \frac{1}{z^4} + \dots$$

$$S(z) = \sum_{s=0}^{\infty} (-1)^s \frac{v_{2s+1}}{z^{2s+1}} = -\frac{3.7}{1!216} \frac{1}{z} + \frac{7.9.11.13.15.19}{3!216^3} \frac{1}{z^3} - \dots$$

Hence we obtain the asymptotic expansions of the Airy functions and of their derivatives (with $\xi = \frac{2}{3}x^{3/2}$)

$$Ai(x) \sim \frac{1}{2\pi^{1/2}x^{1/4}} \mathrm{e}^{-\xi} L(-\xi) \tag{2.45}$$

$$Ai'(x) \sim -\frac{x^{1/4}}{2\pi^{1/2}} \mathrm{e}^{-\xi} M(-\xi) \tag{2.46}$$

$$Bi(x) \sim \frac{1}{\pi^{1/2}x^{1/4}} \mathrm{e}^{\xi} L(\xi) \tag{2.47}$$

$$Bi'(x) \sim \frac{x^{1/4}}{\pi^{1/2}} \mathrm{e}^{\xi} M(\xi) \tag{2.48}$$

$$Ai(-x) \sim \frac{1}{\pi^{1/2}x^{1/4}} \left[\sin\left(\xi - \frac{\pi}{4}\right) Q(\xi) + \cos\left(\xi - \frac{\pi}{4}\right) P(\xi)\right] \tag{2.49}$$

$$Ai'(-x) \sim \frac{x^{1/4}}{\pi^{1/2}} \left[\sin\left(\xi - \frac{\pi}{4}\right) R(\xi) - \cos\left(\xi - \frac{\pi}{4}\right) S(\xi)\right] \tag{2.50}$$

$$Bi(-x) \sim \frac{1}{\pi^{1/2}x^{1/4}} \left[-\sin\left(\xi - \frac{\pi}{4}\right) P(\xi) + \cos\left(\xi - \frac{\pi}{4}\right) Q(\xi)\right] \tag{2.51}$$

$$Bi'(-x) \sim \frac{x^{1/4}}{\pi^{1/2}} \left[\sin\left(\xi - \frac{\pi}{4}\right) S(\xi) + \cos\left(\xi - \frac{\pi}{4}\right) R(\xi)\right]. \tag{2.52}$$

2.1.4.4 *The Stokes phenomenon*

The Stokes phenomenon is linked to the asymptotic series of Airy functions (and other similar functions) which have different form in different sectors of the complex plane. In this subsection, we will briefly describe this "phenomenon" that Stokes discovered in his investigations on the asymptotic expansion of Airy functions [Stokes (1851), (1858)]. Here we follow the analysis by Copson (1963), (1967) of the Stokes phenomenon for the asymptotic expansion of Airy functions.

Equation (2.45) may also be written as

$$Ai(z) \sim \frac{1}{2\pi z^{1/4}} \mathrm{e}^{-\frac{2}{3}z^{3/2}} \sum_{n=0}^{\infty} \frac{\Gamma(3n+\frac{1}{2})}{3^{2n}(2n)!} \frac{(-1)^n}{z^{3n/2}}, \tag{2.53}$$

as $|z| \to \infty$ in the sector $|\mathrm{ph}\,\mathrm{z}| < \pi$. In order to extend the range of values of $\mathrm{ph}\,\mathrm{z}$, we rearrange Eq.(2.3)

$$Ai(z) = -\mathrm{j}Ai(\mathrm{j}z) - \mathrm{j}^2 Ai(\mathrm{j}^2 z).$$

Taking $\mathrm{j} = \mathrm{e}^{2\mathrm{i}\pi/3}$ and $\mathrm{j}^2 = \mathrm{e}^{4\mathrm{i}\pi/3}$, with $-\frac{5}{3}\pi < \mathrm{ph}\,\mathrm{z} < -\frac{1}{3}\pi$, we have $-\pi < \mathrm{ph}\,(\mathrm{jz}) < \frac{1}{3}\pi$ and $-\frac{1}{3}\pi < \mathrm{ph}\,(\mathrm{j^2z}) < \pi$ then for $Ai(\mathrm{j}z)$ and $Ai(\mathrm{j}^2 z)$, we can use the asymptotic expansion given above (Eq. (2.53)). Therefore, we may write

$$Ai(z) \sim F(z) - \mathrm{i}G(z),$$

where

$$F(z) \sim \frac{1}{2\pi z^{1/4}} \mathrm{e}^{-\frac{2}{3}z^{3/2}} \sum_{n=0}^{\infty} \frac{\Gamma(3n+\frac{1}{2})}{3^{2n}(2n)!} \frac{(-1)^n}{z^{3n/2}},$$

and

$$G(z) \sim \frac{1}{2\pi z^{1/4}} \mathrm{e}^{\frac{2}{3}z^{3/2}} \sum_{n=0}^{\infty} \frac{\Gamma(3n+\frac{1}{2})}{3^{2n}(2n)!} \frac{1}{z^{3n/2}},$$

as $|z| \to \infty$ in $-\frac{5}{3}\pi < \mathrm{ph}\,\mathrm{z} < -\frac{1}{3}\pi$.

Now, taking $\mathrm{j} = \mathrm{e}^{-4\mathrm{i}\pi/3}$ and $\mathrm{j}^2 = \mathrm{e}^{-2\mathrm{i}\pi/3}$, with $\frac{1}{3}\pi < \mathrm{ph}\,\mathrm{z} < \frac{5}{3}\pi$, we have $-\pi < \mathrm{ph}\,(\mathrm{jz}) < \frac{1}{3}\pi$ and $-\frac{1}{3}\pi < \mathrm{ph}\,(\mathrm{j}^2\mathrm{z}) < \pi$, then we can again use the above asymptotic expansion. However, the result is now

$$Ai(z) \sim F(z) + \mathrm{i}G(z),$$

as $|z| \to \infty$ in $\frac{1}{3}\pi < \mathrm{ph}\,\mathrm{z} < \frac{5}{3}\pi$.

Consequently we have obtained three asymptotic expansions for $Ai(z)$:

$$\begin{array}{lll} F(z) & \text{when} & -\pi < \mathrm{ph}\,\mathrm{z} < \pi \\ F(z) - \mathrm{i}G(z) & \text{when} & -\dfrac{5}{3}\pi < \mathrm{ph}\,\mathrm{z} < -\dfrac{1}{3}\pi \\ F(z) + \mathrm{i}G(z) & \text{when} & \dfrac{1}{3}\pi < \mathrm{ph}\,\mathrm{z} < \dfrac{5}{3}\pi\,. \end{array}$$

However, as the angle domains overlap, we obtain a paradox which is precisely the Stokes phenomenon. In order to solve this paradox, we consider the point $z_1 = \rho\,\mathrm{e}^{\mathrm{i}(\pi-\alpha)}$ with $\rho > 0$ and $-\frac{2}{3}\pi < \alpha < \frac{2}{3}\pi$, so that $\frac{1}{3}\pi < \mathrm{ph}\,\mathrm{z}_1 < \frac{5}{3}\pi$, and the point $z_2 = \rho\,\mathrm{e}^{-\mathrm{i}(\pi+\alpha)}$ so that $-\frac{5}{3}\pi < \mathrm{ph}\,\mathrm{z}_2 < -\frac{1}{3}\pi$. The Airy function $Ai(z)$ should have the same asymptotic expansions at z_1 and z_2, for it is an entire function. This is indeed the case as we have

$$F(z_2) = \mathrm{i}G(z_1), \qquad G(z_2) = \mathrm{i}F(z_1),$$

so we can write

$$F(z_2) - \mathrm{i}G(z_2) = F(z_1) + \mathrm{i}G(z_1).$$

The asymptotic expansions $F(z)$ and $F(z) + \mathrm{i}G(z)$ are both valid in the sector $\frac{1}{3}\pi < \mathrm{ph}\,\mathrm{z} < \pi$. If we write $z = \rho \mathrm{e}^{\mathrm{i}(\pi-\alpha)}$ where $\rho > 0$ and $0 < \alpha < \frac{2}{3}\pi$ we find that

$$|F(z)| \sim \frac{1}{2\sqrt{\pi}\rho^{1/4}} \mathrm{e}^{\frac{2}{3}\rho^{2/3}\sin\frac{3}{2}\alpha} \quad \text{and} \quad |G(z)| \sim \frac{1}{2\sqrt{\pi}\rho^{1/4}} \mathrm{e}^{-\frac{2}{3}\rho^{2/3}\sin\frac{3}{2}\alpha}.$$

In the chosen sector $\sin\frac{3}{2}\alpha > 0$ and $|G(z)/F(z)|$ tends to zero exponentially as $|z| \to \infty$, therefore in this sector $G(z)$ may be neglected.

In recent years, the study of the Stokes phenomenon turned out to be of prime importance in the understanding of divergent series. For instance, M. Berry (1989-a,b) considered the smoothing of Stokes's discontinuities, leading to the idea of hyperasymptotics and to the phenomenon of resurgence [Berry & Howls (1991); Berry (1992a)].

2.2 Properties of Airy functions

2.2.1 *Zeros of Airy functions*

Zeros of the Airy function $Ai(x)$ are located on the negative part of the real axis. Following the notation of Miller (1946), we define a_s and a'_s, the sth zeros of $Ai(x)$ and $Ai'(x)$, b_s and b'_s the real zeros of $Bi(x)$ and $Bi'(x)$, β_s and β'_s the complex zeros of $Bi(x)$ and $Bi'(x)$ in the region defined by $\frac{\pi}{3} < \arg(x) < \frac{\pi}{2}$. The complex zeros of $Bi(x)$ and $Bi'(x)$ in the region $-\frac{\pi}{2} < \arg(x) < -\frac{\pi}{3}$ are the conjugates of β_s and β'_s. We thus obtain

$$a_s = -f\left[\frac{3\pi}{8}(4s-1)\right] \tag{2.54}$$

$$a'_s = -g\left[\frac{3\pi}{8}(4s-3)\right] \tag{2.55}$$

$$b_s = -f\left[\frac{3\pi}{8}(4s-3)\right] \tag{2.56}$$

$$b'_s = -g\left[\frac{3\pi}{8}(4s-1)\right] \tag{2.57}$$

$$\beta_s = \mathrm{e}^{\mathrm{i}\pi/3} f\left[\frac{3\pi}{8}(4s-1) + \frac{3\mathrm{i}}{4}\ln(2)\right] \tag{2.58}$$

$$\beta'_s = \mathrm{e}^{\mathrm{i}\pi/3} g\left[\frac{3\pi}{8}(4s-3) + \frac{3\mathrm{i}}{4}\ln(2)\right]. \tag{2.59}$$

We also have the relations:

$$Ai'(a_s) = (-1)^{s-1} f_1\left[\frac{3\pi}{8}(4s-1)\right] \tag{2.60}$$

$$Ai(a'_s) = (-1)^{s-1} g_1\left[\frac{3\pi}{8}(4s-3)\right] \tag{2.61}$$

$$Bi'(b_s) = (-1)^{s-1} f_1 \left[\frac{3\pi}{8}(4s-3)\right] \tag{2.62}$$

$$Bi(b'_s) = (-1)^s g_1 \left[\frac{3\pi}{8}(4s-1)\right] \tag{2.63}$$

$$Bi'(\beta_s) = (-1)^s \sqrt{2} \mathrm{e}^{-\mathrm{i}\pi/6} f_1 \left[\frac{3\pi}{8}(4s-1) + \frac{3\mathrm{i}}{4}\ln(2)\right] \tag{2.64}$$

$$Bi(\beta_s) = (-1)^{s-1} \sqrt{2} \mathrm{e}^{\mathrm{i}\pi/6} g_1 \left[\frac{3\pi}{8}(4s-3) + \frac{3\mathrm{i}}{4}\ln(2)\right]. \tag{2.65}$$

The distribution of the zeros of Ai and Bi in the complex plane are given in Figs. (2.4) and (2.5) respectively.

The functions $f(x)$, $g(x)$, $f_1(x)$ and $g_1(x)$ are defined, with $|x| \gg 1$, by the relations

$$f(x) \approx x^{2/3}\left(1 + \frac{5}{48x^2} - \frac{5}{36x^4} + \frac{77\,125}{82\,944x^6} - \ldots\right) \tag{2.66}$$

$$g(x) \approx x^{2/3}\left(1 - \frac{7}{48x^2} + \frac{35}{288x^4} - \frac{181\,223}{207\,360x^6} + \ldots\right) \tag{2.67}$$

$$f_1(x) \approx \frac{x^{1/6}}{\pi^{1/2}}\left(1 + \frac{5}{48x^2} - \frac{1\,525}{4\,608x^4} + \frac{2\,397\,875}{663\,552x^6} - \ldots\right) \tag{2.68}$$

$$g_1(x) \approx \frac{x^{-1/6}}{\pi^{1/2}}\left(1 - \frac{7}{96x^2} + \frac{1\,673}{6\,144x^4} - \frac{84\,394\,709}{26\,552\,080x^6} + \ldots\right). \tag{2.69}$$

2.2.2 *The spectral zeta function*

As the Airy function Ai is an entire function, we can make use of the Weierstrass infinite product

$$Ai(z) = Ai(0)\,\mathrm{e}^{-\kappa z} \prod_{n=1}^{\infty}\left(1 + \frac{z}{|a_n|}\right)\mathrm{e}^{-z/|a_n|}, \tag{2.70}$$

where a_n are the zeros of the Airy function Ai and

$$\kappa = \left|\frac{Ai'(0)}{Ai(0)}\right| = \frac{3^{5/6}\,\Gamma(2/3)^2}{2\pi} = 0.72901113\ldots.$$

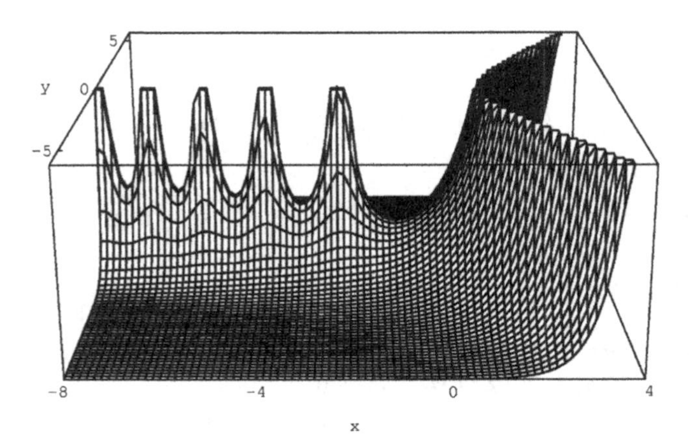

Fig. 2.4 Plot of $1/|Ai\,(x+\mathrm{i}y)|$. Zeros of $|Ai\,(x+\mathrm{i}y)|$ are located on the negative part of the real axis. The modulus of the Airy function $|Ai(z)|$ blows up outside this axis, except in the sector defined by $-\frac{\pi}{3} < \arg(z) < \frac{\pi}{3}$ where it goes to 0.

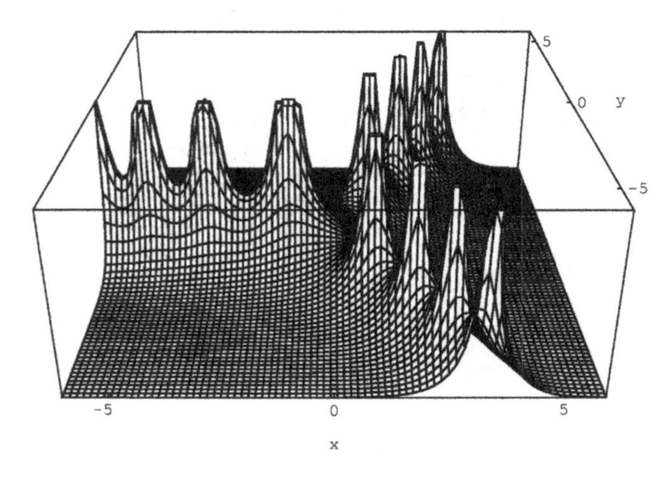

Fig. 2.5 Plot of $1/|Bi\,(x+\mathrm{i}y)|$. The real zeros of $|Bi\,(x+\mathrm{i}y)|$ can be discerned on the negative part of the real axis, and the conjugated complex pair of zeros in the sectors $\frac{\pi}{3} < \arg(z) < \frac{\pi}{2}$ and $-\frac{\pi}{2} < \arg(z) < -\frac{\pi}{3}$.

We now take the logarithmic derivative of this product

$$\frac{\mathrm{d}}{\mathrm{d}z}\ln Ai(z) = \frac{Ai'(z)}{Ai(z)} = -\kappa + \sum_{n=1}^{\infty} \frac{1}{z+|a_n|} - \frac{1}{|a_n|}. \tag{2.71}$$

One more differentiation of this result provides

$$\frac{\mathrm{d}^2}{\mathrm{d}z^2}\ln Ai(z) = z - \left(\frac{Ai'(z)}{Ai(z)}\right)^2 = -\sum_{n=1}^{\infty}\frac{1}{(z+|a_n|)^2}. \tag{2.72}$$

Therefore, we obtain the following properties of the zeros of the Airy function

$$\sum_{n=1}^{\infty}\frac{1}{a_n^2} = \kappa^2. \tag{2.73}$$

More generally, we may define the Airy zeta function constructed with the zeros of this function

$$Z^-(s) = \sum_{n=1}^{\infty}\frac{1}{|a_n|^s}. \tag{2.74}$$

According to the asymptotic property of the zeros a_n (2.54), (2.66) and (2.67), the convergence of this series is ensured for $s > \frac{3}{2}$. Other particular cases of this zeta function may be obtained by taking the successive derivatives of Eq. (2.71). For instance, we have

$$Z^-(3) = \sum_{n=1}^{\infty}\frac{1}{|a_n|^3} = \frac{1}{2} - \kappa^3. \tag{2.75}$$

The same can be done for the derivative of the Airy function ($Ai''(0) = 0$)

$$Ai'(z) = Ai'(0)\prod_{n=1}^{\infty}\left(1 + \frac{z}{|a'_n|}\right)\mathrm{e}^{-z/|a'_n|}, \tag{2.76}$$

for which we define the zeta function[1] ($s > \frac{3}{2}$)

$$Z^+(s) = \sum_{n=1}^{\infty}\frac{1}{|a'_n|^s}. \tag{2.77}$$

Some particular cases are given in Table (2.1) [Crandal (1996); Voros (1999)]. The values for $n = 1$ are obtained by analytic continuations.

In this table, we can see the noteworthy result: *the sum of the inverse cubes of the zeros of $Ai'(z)$ is unity.*

Flajolet & Louchard (2001) studied the area Airy distribution which occurs in a number of combinatorial structures, such as path length in trees, area below random walks, displacement in parking sequence,.... They found interesting results for the spectral function $Z^-(s)$ which is

[1]The notation: $Z^{\pm}(s)$ comes from Voros (1999).

Table 2.1 Some particular values of the Airy zeta functions $Z^{\pm}(s)$.

n	1	2	3	4
$Z^-(n)$	$-\kappa$	κ^2	$\frac{1}{2}-\kappa^3$	$\frac{1}{3}\kappa-\kappa^4$
$Z^+(n)$	0	$\frac{1}{\kappa}$	1	$\frac{1}{\kappa^2}$

closely related to this area distribution. In particular, they extended the definition Eq. (2.74) by analytic continuation, and using the Mellin transform

$$\mathcal{M}\left[\frac{Ai'(z)}{Ai(z)}+\kappa\right](s)=Z^-(1-s)\,\frac{\pi}{\sin\,\pi s},$$

where $\mathcal{M}$ represents a Mellin transform, they showed that

$$Z^-(-1)=Z^-(-2)=Z^-(-4)=\ldots=0.$$

Other results may be found in the paper by Flajolet & Louchard (2001).

2.2.3 *Inequalities*

Some of the previous results may be used to prove inequalities for the Airy function Ai, which is concave in the domain $(-a_1,\infty)$ [Salmassi (1999)]

$$x\,Ai^2(x)\le Ai'^{\,2}(x). \tag{2.78}$$

For two different arguments

$$Ai(x)Ai(y)\le Ai^2\left(\frac{x+y}{2}\right), \tag{2.79}$$

and for the scaling

$$Ai(x)^{\alpha}Ai(0)^{1-\alpha}\le Ai(\alpha x);\quad 0\le\alpha\le 1. \tag{2.80}$$

Similar inequalities may be found for the function Bi.

2.2.4 *Connection with Bessel functions*

As mentioned in the introduction, Airy functions Ai and Bi may be alternatively written in terms of Bessel functions I and J of order $1/3$, and of order $2/3$ for their derivatives, $\xi = \frac{2}{3}x^{3/2}$ [Jeffreys (1942); Olver (1974)]

$$Ai(x) = \frac{\sqrt{x}}{3}\left[I_{-1/3}(\xi) - I_{1/3}(\xi)\right] \tag{2.81}$$
$$= \frac{1}{\pi}\sqrt{\frac{x}{3}}K_{1/3}(\xi)$$

$$Ai(-x) = \frac{\sqrt{x}}{3}\left[J_{-1/3}(\xi) + J_{1/3}(\xi)\right] \tag{2.82}$$
$$= \sqrt{\frac{x}{3}}\,\Re\left[\mathrm{e}^{\mathrm{i}\frac{\pi}{6}}H^{(1)}_{1/3}(\xi)\right]$$

$$Ai'(x) = -\frac{x}{3}\left[I_{-2/3}(\xi) - I_{2/3}(\xi)\right] \tag{2.83}$$
$$= -\frac{1}{\pi}\frac{x}{\sqrt{3}}K_{2/3}(\xi)$$

$$Ai'(-x) = -\frac{x}{3}\left[J_{-2/3}(\xi) - J_{2/3}(\xi)\right] \tag{2.84}$$
$$= \frac{x}{\sqrt{3}}\,\Re\left[\mathrm{e}^{-\mathrm{i}\frac{\pi}{6}}H^{(1)}_{2/3}(\xi)\right]$$

$$Bi(x) = \sqrt{\frac{x}{3}}\left[I_{-1/3}(\xi) + I_{1/3}(\xi)\right] \tag{2.85}$$
$$= \sqrt{\frac{x}{3}}\,\Re\left[\mathrm{e}^{\mathrm{i}\frac{\pi}{6}}H^{(1)}_{1/3}(-\mathrm{i}\xi)\right]$$

$$Bi(-x) = \sqrt{\frac{x}{3}}\left[J_{-1/3}(\xi) - J_{1/3}(\xi)\right] \tag{2.86}$$
$$= -\sqrt{\frac{x}{3}}\,\Im\left[\mathrm{e}^{\mathrm{i}\frac{\pi}{6}}H^{(1)}_{1/3}(\xi)\right]$$

$$Bi'(x) = \frac{x}{\sqrt{3}}\left[I_{-2/3}(\xi) + I_{2/3}(\xi)\right] \tag{2.87}$$
$$= \frac{x}{\sqrt{3}}\,\Re\left[\mathrm{e}^{\mathrm{i}\frac{\pi}{6}}H^{(1)}_{2/3}(-\mathrm{i}\xi)\right]$$

$$Bi'(-x) = \frac{x}{\sqrt{3}}\left[J_{-2/3}(\xi) + J_{2/3}(\xi)\right] \tag{2.88}$$
$$= -\frac{x}{\sqrt{3}}\,\Im\left[\mathrm{e}^{-\mathrm{i}\frac{\pi}{6}}H^{(1)}_{2/3}(\xi)\right].$$

The modified Bessel function $K_\nu(z)$ is defined by

$$K_\nu(z) = \frac{\pi}{2}\frac{I_{-\nu}(z) - I_\nu(z)}{\sin(\pi\nu)},$$

and the Hankel functions $H_\nu^{(1)}(z)$ and $H_\nu^{(2)}(z)$, by

$$H_\nu^{(1)}(z) = J_\nu(z) + \mathrm{i}Y_\nu(z),$$

$$H_\nu^{(2)}(z) = J_\nu(z) - \mathrm{i}Y_\nu(z),$$

with the Weber function

$$Y_\nu(z) = \frac{J_\nu(z) - J_{-\nu}(z)}{\sin(\pi\nu)}.$$

2.2.5 *Modulus and phase of Airy functions*

2.2.5.1 *Definitions*

We define the modulus $M(x)$ and the phase $\theta(x)$ of the functions $Ai(x)$ and $Bi(x)$, and the modulus $N(x)$ and the phase $\phi(x)$ of the functions $Ai'(x)$ and $Bi'(x)$, for any $x > 0$ by the relations [Miller (1946); Abramowitz & Stegun, (1965)]

$$Ai(-x) = M(x)\cos\left[\theta(x)\right] \tag{2.89}$$

$$Bi(-x) = M(x)\sin\left[\theta(x)\right] \tag{2.90}$$

$$Ai'(-x) = N(x)\cos\left[\phi(x)\right] \tag{2.91}$$

$$Bi'(-x) = N(x)\sin\left[\phi(x)\right], \tag{2.92}$$

and the inverse relations

$$M(x) = \left[Ai^2(-x) + Bi^2(-x)\right]^{1/2} \tag{2.93}$$

$$\theta(x) = \arctan\left[\frac{Bi(-x)}{Ai(-x)}\right] \tag{2.94}$$

$$N(x) = \left[Ai'^2(-x) + Bi'^2(-x)\right]^{1/2} \tag{2.95}$$

$$\phi(x) = \arctan\left[\frac{Bi'(-x)}{Ai'(-x)}\right]. \tag{2.96}$$

2.2.5.2 *Differential equations*

The moduli and phases are solutions to the following differential equations, for $x > 0$ [Miller (1946); Abramowitz & Stegun (1965)]

$$M^2\theta' = -\frac{1}{\pi} \tag{2.97}$$

$$N^2\phi' = -\frac{x}{\pi} \tag{2.98}$$

$$N^2 = M'^2 + M^2\theta'^2 \tag{2.99}$$

$$NN' = -xMM' \tag{2.100}$$

$$\tan(\phi - \theta) = \frac{M\theta'}{M'} \tag{2.101}$$

$$\sin(\phi - \theta) = \frac{1}{\pi MN} \tag{2.102}$$

$$M'' + xM - \frac{1}{\pi^2 M^3} = 0 \tag{2.103}$$

$$\left(M^2\right)''' + 4x\left(M^2\right)' + 2M^2 = 0 \tag{2.104}$$

$$\theta'^2 + \frac{1}{2}\frac{\theta'''}{\theta'} - \frac{3}{4}\left(\frac{\theta''}{\theta'}\right)^2 = x. \tag{2.105}$$

This last expression may be alternatively written $\theta'^2 + \frac{1}{2}\{\theta, x\} = x$, where $\{\theta, x\}$ is the Schwarzian derivative of θ with respect to x, as we shall see in §5.2.

2.2.5.3 *Asymptotic expansions*

For $x \gg 1$, the asymptotic series for the moduli and phases of Airy functions are [Miller (1946); Abramowitz & Stegun (1965)]

$$\begin{aligned} M^2(x) &\approx \frac{1}{\pi x^{1/2}} \sum_{k=0}^{\infty} \frac{(-1)^k}{12^k k!} 2^{3k} \left(\frac{1}{2}\right)_{3k} (2x)^{-3k} \\ &\approx \frac{1}{\pi x^{1/2}} \left(1 - \frac{1.3.5}{1!96x^3} + \frac{1.3.5.7.9.11}{2!96^2x^6} \right. \\ &\qquad \left. - \frac{1.3.5.7.9.11.13.15.17}{3!96^3x^9} + \ldots\right) \end{aligned} \tag{2.106}$$

$$\begin{aligned} N^2(x) &\approx \frac{x^{1/2}}{\pi} \sum_{k=0}^{\infty} \frac{(-1)^{k+1}}{12^k k!} \frac{6k+1}{6k-1} 2^{3k} \left(\frac{1}{2}\right)_{3k} (2x)^{-3k} \\ &\approx \frac{x^1/2}{\pi} \left(1 + \frac{1.3}{1!96}\frac{7}{x^3} - \frac{1.3.5.7.9}{2!96^2}\frac{13}{x^6} \right. \\ &\qquad \left. + \frac{1.3.5.7.9.11.13.15}{3!96^3}\frac{19}{x^9} - \ldots\right) \end{aligned} \tag{2.107}$$

where $\left(\frac{1}{2}\right)_{3k}$ is the Pochhammer symbol defined in §2.1.4 (formula (2.40)), and

$$\theta(x) \approx \frac{\pi}{4} - \frac{2}{3}x^{3/2}\left[1 - \frac{5}{4}\frac{1}{(2x)^3} + \frac{1\,105}{96}\frac{1}{(2x)^6} - \frac{82\,825}{128}\frac{1}{(2x)^9} + \frac{1\,282\,031\,525}{14\,336}\frac{1}{(2x)^{12}} - \dots\right] \tag{2.108}$$

$$\phi(x) \approx \frac{3\pi}{4} - \frac{2}{3}x^{3/2}\left[1 + \frac{7}{4}\frac{1}{(2x)^3} - \frac{1\,463}{96}\frac{1}{(2x)^6} + \frac{495\,271}{640}\frac{1}{(2x)^9} - \frac{206\,530\,429}{2\,048}\frac{1}{(2x)^{12}} - \dots\right]. \tag{2.109}$$

2.2.5.4 *Functions of positive arguments*

When the argument of the Airy functions is positive, they do not oscillate, but increase or decrease exponentially. It is then convenient to "exponentially" normalise these functions [Alexander & Manolopoulos (1987)]

$$ai(x) = \mathrm{e}^{\xi} Ai(x) \tag{2.110}$$

$$ai'(x) = \mathrm{e}^{\xi} Ai'(x) \tag{2.111}$$

$$bi(x) = \mathrm{e}^{-\xi} Bi(x) \tag{2.112}$$

$$bi'(x) = \mathrm{e}^{-\xi} Bi'(x) \tag{2.113}$$

with $\xi = \frac{2}{3}x^{3/2}$, $x > 0$. We can then, as above, define the moduli $\overline{M}(x)$ and $\overline{N}(x)$, and the hyperbolic phases $\chi(x)$ and $\eta(x)$ of Airy functions by

$$ai(x) = \overline{M}(x)\cosh\left[\chi(x)\right] \tag{2.114}$$

$$bi(x) = \overline{M}(x)\sinh\left[\chi(x)\right] \tag{2.115}$$

$$ai'(x) = \overline{N}(x)\cosh\left[\eta(x)\right] \tag{2.116}$$

$$bi'(x) = \overline{N}(x)\sinh\left[\eta(x)\right], \tag{2.117}$$

and the inverse relations

$$\overline{M}(x) = \left[bi^2(x) - ai^2(x)\right]^{1/2} \tag{2.118}$$

$$\chi(x) = \operatorname{arctanh}\left[\frac{ai(x)}{bi(x)}\right] \tag{2.119}$$

$$\overline{N}(x) = \left[bi'^2(x) - ai'^2(x)\right]^{1/2} \tag{2.120}$$

$$\eta(x) = \operatorname{arctanh}\left[\frac{ai'(x)}{bi'(x)}\right]. \tag{2.121}$$

We can also calculate the asymptotic expansion of these functions. For $x \gg 1$, we have

$$\overline{M}^2(x) \approx \frac{3}{4\pi x^{1/2}} \left[1 + \frac{25}{72x^{3/2}} + \dots\right] \quad (2.122)$$

$$\overline{N}^2(x) \approx \frac{3x^{1/2}}{4\pi} \left[1 + \frac{7}{24x^{3/2}} + \dots\right] \quad (2.123)$$

$$\chi(x) \approx \frac{1}{2}\ln 3 - \frac{5}{36x^{3/2}} + \dots \quad (2.124)$$

$$\eta(x) \approx -\frac{1}{2}\ln 3 + \dots . \quad (2.125)$$

2.3 Inhomogeneous Airy functions

2.3.1 *Definitions*

In this section, we consider the inhomogeneous differential equation of the second order [Scorer (1950); Abramowitz & Stegun (1965)]

$$y'' - xy = \pm\pi^{-1}, \quad (2.126)$$

the resolution of which is done by a similar method to the homogeneous one. The solutions are the inhomogeneous functions $Gi(x)$ and $Hi(x)$ (also called *Scorer functions*), according to the sign of the right-hand side $-$ or $+$ respectively. The integral representations of these functions are given by

$$Gi(x) = \frac{1}{\pi}\int_0^{\infty} \sin\left(\frac{t^3}{3} + xt\right)\mathrm{d}t \quad (2.127)$$

and

$$Hi(x) = \frac{1}{\pi}\int_0^{\infty} \mathrm{e}^{-t^3/3+xt}\mathrm{d}t. \quad (2.128)$$

We can alternatively define these functions from the primitives of the homogeneous functions [Scorer (1950); Olver (1974)]

$$Gi(x) = Bi(x)\int_x^{+\infty} Ai(t)\mathrm{d}t + Ai(x)\int_0^x Bi(t)\mathrm{d}t \quad (2.129)$$

$$Hi(x) = Bi(x)\int_{-\infty}^x Ai(t)\mathrm{d}t - Ai(x)\int_{-\infty}^x Bi(t)\mathrm{d}t. \quad (2.130)$$

These functions are related to the homogeneous Airy function $Bi(x)$ by the important relation

$$Bi(x) = Gi(x) + Hi(x). \tag{2.131}$$

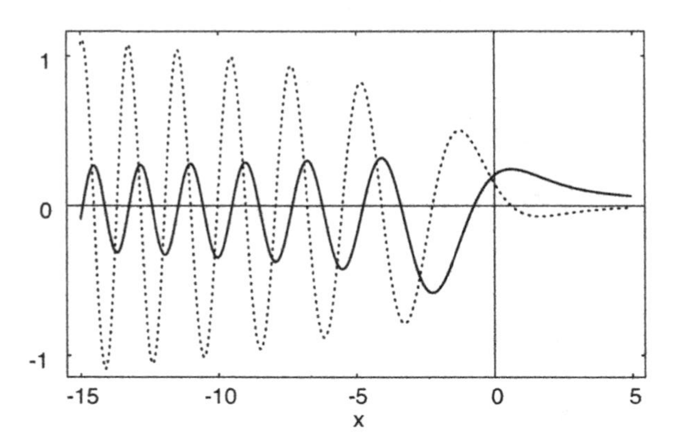

Fig. 2.6 Plot of the inhomogeneous Airy function Gi (solid line) and its derivative Gi' (dotted line).

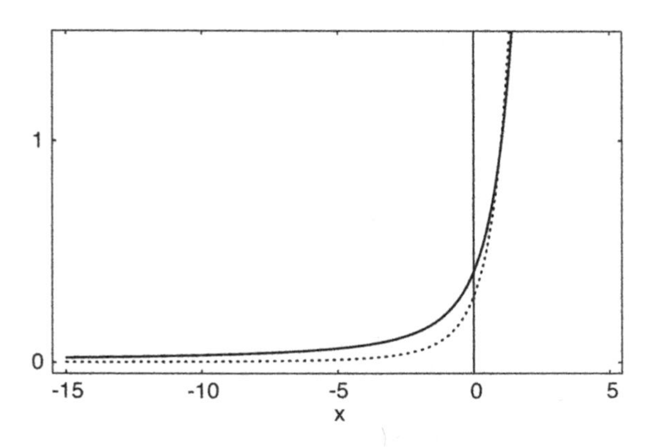

Fig. 2.7 Plot of the inhomogeneous Airy function Hi (solid line) and its derivative Hi' (dotted line).

We have plotted on Figs. (2.6) and (2.7) the inhomogeneous Airy functions $Gi(x)$, $Hi(x)$ and their derivatives $Gi'(x)$ and $Hi'(x)$.

The general solutions of Eq. (2.126) are then

$$y(x) = aAi(x) + bBi(x) + Gi(x), \tag{2.132}$$

or

$$y(x) = cAi(x) + dBi(x) + Hi(x), \tag{2.133}$$

according to Eq. (2.131); a, b, c and d being constants of integration.

2.3.2 *Properties of inhomogeneous Airy functions*

2.3.2.1 *Values at the origin*

The values at the origin of inhomogeneous Airy functions, and of their derivatives, are given by [Scorer (1950); Gordon (1970)]

$$Gi\,(0) = \frac{Hi\,(0)}{2} = \frac{Ai\,(0)}{\sqrt{3}} = \frac{1}{3^{7/6}\Gamma\left(\frac{2}{3}\right)}, \tag{2.134}$$

$$Gi'\,(0) = \frac{Hi'\,(0)}{2} = -\frac{Ai'\,(0)}{\sqrt{3}} = \frac{1}{3^{5/6}\Gamma\left(\frac{1}{3}\right)}. \tag{2.135}$$

2.3.2.2 *Other integral representations*

Besides the definitions (2.127) and (2.128) of Gi and Hi, these functions may be given in terms of the following integrals, for $x > 0$ [Gordon (1970)]

$$Gi(x) = \frac{4x^2}{3\sqrt{3}\pi^2}\int_0^\infty \frac{K_{1/3}(t)}{\xi^2 - t^2}\mathrm{d}t, \tag{2.136}$$

$$Hi(-x) = \frac{4x^2}{3\sqrt{3}\pi^2}\int_0^\infty \frac{K_{1/3}(t)}{\xi^2 + t^2}\mathrm{d}t, \tag{2.137}$$

where $K_{1/3}(t)$ is the modified Bessel function and $\xi = \frac{2}{3}x^{3/2}$. We also have the following integral representation for $Gi(x)$ [Lee (1980)]:

$$Gi(x) = -\frac{1}{\pi}\int_0^\infty \mathrm{e}^{-t^3/3 - tx/2}\cos\left(\frac{\sqrt{3}}{2}tx + \frac{2\pi}{3}\right)\mathrm{d}t. \tag{2.138}$$

It should be noted that the function $Hi(x)$ is a particular case of the Faxén integral (1921),

$$Fi\,(\alpha, \beta; y) = \int_0^\infty e^{t^\alpha y - t}t^{\beta-1}\mathrm{d}t, \; 0 \leq \Re(\alpha) < 1, \; \Re(\beta) > 0,$$

that is to say

$$Hi(x) = \frac{1}{3^{2/3}\pi}Fi\left(\frac{1}{3}, \frac{1}{3}; 3^{1/3}x\right). \tag{2.139}$$

2.3.3 *Ascending series and asymptotic expansion*

2.3.3.1 *Ascending series*

We can write the integral representation of the function $Hi(x)$ (formula (2.128)) [Scorer (1950); Lee (1980)], as

$$Hi(x) = \frac{1}{\pi} \sum_{k=0}^{\infty} \frac{x^k}{k!} \int_0^{\infty} t^k \mathrm{e}^{-t^3/3} \mathrm{d}t.$$

We recognise, in this relation, the expression of the gamma function, $\Gamma(z) = \int_0^{\infty} \mathrm{e}^{-u} u^{z-1} \mathrm{d}u$, from which we deduce the ascending series

$$Hi(x) = \frac{1}{\pi} \sum_{k=0}^{\infty} 3^{(k-2)/3} \Gamma\left(\frac{k+1}{3}\right) \frac{x^k}{k!}. \tag{2.140}$$

This result may alternatively be written

$$Hi(x) = c_3 f(x) + c_4 g(x) + \frac{1}{\pi} h(x), \tag{2.141}$$

with: $c_3 = Hi(0)$, $c_4 = Hi'(0)$, and the series

$$f(x) = 1 + \frac{1}{3!}x^3 + \frac{1.4}{6!}x^6 + \frac{1.4.7}{9!}x^9 + \dots$$

$$g(x) = x\left(1 + \frac{2}{4!}x^3 + \frac{2.5}{7!}x^6 + \frac{2.5.8}{10!}x^9 + \dots\right)$$

$$h(x) = x^2\left(\frac{1}{2} + \frac{3}{5!}x^3 + \frac{3.6}{8!}x^6 + \frac{3.6.9}{11!}x^9 + \dots\right).$$

Note that the series of f and g are identical to the series defined in §2.1.4.2. The ascending series of $Gi(x)$ can be deduced from those of $Hi(x)$ and $Bi(x)$ (formula (2.39)) thanks to the relation (2.131). We obtain

$$Gi(x) = \frac{1}{3} Bi(x) - \frac{1}{\pi} h(x). \tag{2.142}$$

For the derivatives, we have

$$Hi'(x) = \frac{1}{\pi} \sum_{k=0}^{\infty} 3^{(k-1)/3} \Gamma\left(\frac{k+2}{3}\right) \frac{x^k}{k!}. \tag{2.143}$$

The ascending series of $Gi'(x)$ is deduced (as for $Gi(x)$) thanks to the ascending series of $Bi'(x)$ and $Hi'(x)$.

2.3.3.2 *Asymptotic expansions*

The asymptotic expansions of the functions $Gi(x)$ and $Hi(-x)$ are, for $x \gg 1$ [Scorer (1950); Lee (1980)]:

$$Gi(x) \approx \frac{1}{\pi x} \sum_{n=0}^{\infty} \frac{(3n)!}{3^n n!} \frac{1}{x^{3n}} \tag{2.144}$$

$$\approx \frac{1}{\pi x} \left(1 + \frac{2!}{x^3} + \frac{5!}{3x^6} + \frac{8!}{3.6x^9} + \dots \right)$$

$$Hi(-x) \approx \frac{1}{\pi x} \sum_{n=0}^{\infty} \frac{(-1)^n (3n)!}{3^n n!} \frac{1}{x^{3n}} . \tag{2.145}$$

$$\approx \frac{1}{\pi x} \left(1 - \frac{2!}{x^3} + \frac{5!}{3x^6} - \frac{8!}{3.6x^9} + \dots \right) .$$

From the previous two formulae, and from the relation (2.131), we can obtain the expansion of $Gi(-x)$ and $Hi(x)$. So we have

$$Gi(-x) = Bi(-x) - Hi(-x),$$

and

$$Hi(x) = Bi(x) - Gi(x),$$

with the asymptotic expansions of $Bi(x)$ and $Bi(-x)$ being given by the formulae (2.47) and (2.51). Olver (1954) gives the asymptotic expansions under an equivalent form

$$Gi(x) \approx \frac{1}{\pi x} \left[1 + \frac{1}{x^3} \sum_{s=0}^{\infty} \frac{(3s+2)!}{s!\,(3x^3)^s} \right] , \quad x \to +\infty \tag{2.146}$$

$$Hi(x) \approx -\frac{1}{\pi x} \left[1 + \frac{1}{x^3} \sum_{s=0}^{\infty} \frac{(3s+2)!}{s!\,(3x^3)^s} \right] , \quad x \to -\infty. \tag{2.147}$$

2.3.4 *Zeros of the Scorer functions*

In an interesting paper on the zeros of the Scorer functions, Gil, Segura and Temme (2003) gave several important results on the subject. Here, we limit ourselves to two:

- The Scorer function Hi has no real zero, but infinitely many complex zeros on the half line $\mathrm{ph}\, z = \pi/3$, and at the complex conjugated values.
- The derivative Gi' has exactly one positive zero at $g' = 0.609075417\dots$, as can be seen on Fig. (2.6).

For the other values and properties of these zeros, we refer the reader to the paper by Gil *et al.* (2003).

2.4 Squares and products of Airy functions

2.4.1 *Differential equation and integral representation*

The homogeneous differential equation of the third order

$$y''' - 4xy' - 2y = 0 \tag{2.148}$$

has three linearly independent solutions [Aspnes (1966); Reid (1995)]: $Ai^2(x)$, $Ai(x)Bi(x)$ and $Bi^2(x)$, whose Wronskian is

$$W\left\{Ai^2(x), Ai(x)Bi(x), Bi^2(x)\right\} = \frac{2}{\pi^3}\,. \tag{2.149}$$

The solution $Ai^2(x)$ may be written as the integral

$$Ai^2(x) = \frac{1}{4\pi^2}\iint\limits_{\mathbb{R}^2} \mathrm{e}^{\mathrm{i}\left[u^3/3+v^3/3+(u+v)x\right]}\mathrm{d}u\mathrm{d}v.$$

Changing the variables

$$s = \frac{1}{2}(v-u)\,; \quad t = v+u,$$

and integrating with respect to the variable s, we obtain an integral representation of $Ai^2(x)$,

$$Ai^2(x) = \frac{1}{2\pi^{3/2}}\int_0^\infty \cos\left(\frac{t^3}{12} + tx + \frac{\pi}{4}\right)\frac{\mathrm{d}t}{\sqrt{t}}. \tag{2.150}$$

Similarly, we have an integral representation of $Ai(x)Bi(x)$

$$Ai(x)Bi(x) = \frac{1}{2\pi^{3/2}}\int_0^\infty \sin\left(\frac{t^3}{12} + tx + \frac{\pi}{4}\right)\frac{\mathrm{d}t}{\sqrt{t}}, \tag{2.151}$$

while an interesting formula for $Ai^2(x)$ is given by

$$Ai^2(x) = \frac{1}{4\pi\sqrt{3}}\int_0^\infty J_0\left(\frac{t^3}{12} + tx\right)t\,\mathrm{d}t, \qquad x \geq 0. \tag{2.152}$$

We can generalise the preceding result in the case of the product of Airy functions with different arguments [Vallée *et al.* (1997)]:

$$\begin{aligned} &Ai(u)Ai(v) \\ &= \frac{1}{2^{1/3}\pi}\int_{-\infty}^{+\infty} Ai\left[2^{2/3}\left(t^2 + \frac{u+v}{2}\right)\right]\mathrm{e}^{\mathrm{i}(u-v)t}\mathrm{d}t \\ &= \frac{1}{2\pi^{3/2}}\int_0^\infty \cos\left(\frac{t^3}{12} + \frac{u+v}{2}t - \frac{(u-v)^2}{4t} + \frac{\pi}{4}\right)\frac{\mathrm{d}t}{\sqrt{t}}. \end{aligned} \tag{2.153}$$

This result yields

$$Ai^2(x) = \frac{1}{2^{1/3}\pi} \int_{-\infty}^{+\infty} Ai\left[2^{2/3}\left(t^2 + x\right)\right] \mathrm{d}t \tag{2.154}$$

$$Ai(x)Ai(-x) = \frac{1}{2^{1/3}\pi} \int_{-\infty}^{+\infty} Ai\left[2^{2/3}t^2\right] \mathrm{e}^{2\mathrm{i}xt}\mathrm{d}t \tag{2.155}$$

$$Ai(x)Ai(x^*) = \left|Ai(x)\right|^2 \tag{2.156}$$

$$= \frac{1}{2^{1/3}\pi} \int_{-\infty}^{+\infty} Ai\left[2^{2/3}\left(t^2 + \Re(x)\right)\right] \cosh\left[2t\ \Im(x)\right] \mathrm{d}t.$$

The following relations can also be established [Aspnes (1966): cf. §3.5.2]

$$Ai^2(x) = \frac{1}{2^{2/3}\pi} \int_0^{\infty} Ai\left(2^{2/3}x + t\right) \frac{\mathrm{d}t}{\sqrt{t}} \tag{2.157}$$

$$Ai(x)Bi(x) = \frac{1}{2^{2/3}\pi} \int_0^{\infty} Ai\left(2^{2/3}x - t\right) \frac{\mathrm{d}t}{\sqrt{t}} \tag{2.158}$$

$$Ai(x)Ai'(x) = \frac{1}{2\pi} \int_0^{\infty} Ai'\left(2^{2/3}x + t\right) \frac{\mathrm{d}t}{\sqrt{t}}, \tag{2.159}$$

and [Aspnes (1967): cf. §3.5.2]

$$Ai^2(x) = -\frac{1}{2^{2/3}\pi} \int_0^{\infty} Gi\left(2^{2/3}x - t\right) \frac{\mathrm{d}t}{\sqrt{t}} \tag{2.160}$$

$$Ai(x)Bi(x) = \frac{1}{2^{2/3}\pi} \int_0^{\infty} Gi\left(2^{2/3}x + t\right) \frac{\mathrm{d}t}{\sqrt{t}}. \tag{2.161}$$

Finally, Reid (1995) gives the relations

$$Ai^2(x) + Bi^2(x) = \frac{1}{\pi^{3/2}} \int_0^\infty \mathrm{e}^{xt - t^3/12} \frac{\mathrm{d}t}{\sqrt{t}} \tag{2.162}$$

$$Ai\left(x\mathrm{e}^{-\mathrm{i}\pi/6}\right) Ai\left(x\mathrm{e}^{\mathrm{i}\pi/6}\right) = \frac{1}{4\pi^{3/2}} \int_0^{+\infty} \mathrm{e}^{-t^3/12 - x^2/t} \frac{\mathrm{d}t}{\sqrt{t}} \tag{2.163}$$

$$\begin{aligned} Ai(x)Ai(y) &= \frac{\sqrt{3}}{2\pi} \int_0^\infty \mathrm{e}^{-(x^3+y^3)\frac{t^3}{3} - \frac{1}{3t^3}} Ai\left(xyt^2\right) \frac{\mathrm{d}t}{t^2} \\ &= \frac{\sqrt{3}}{2\pi} \int_0^\infty \mathrm{e}^{-\frac{t^3}{3} - \frac{x^3+y^3}{3t^3}} Ai\left(\frac{xy}{t^2}\right) \mathrm{d}t, \;\; x+y>0. \end{aligned} \tag{2.164}$$

2.4.2 *A remarkable identity*

Similarly to the relations in the complex plane (cf. Eqs. (2.3) and (2.5)), Voros (1999) proved the noteworthy result

$$D_0^2 + D_1^2 + D_2^2 - 2(D_0D_1 + D_1D_2 + D_2D_0) + 4 = 0, \tag{2.165}$$

where

$$D_0 = -2\pi\left(Ai^2(z)\right)', \quad D_1 = -2\pi\left(Ai^2(\mathrm{j}^2 z)\right)', \quad D_2 = -2\pi\left(Ai^2(\mathrm{j}z)\right)'.$$

2.4.3 *The product $Ai(x)Ai(-x)$: Airy wavelets*

As we shall see in §4.2, the Airy function allows the definition of a semi-group of transformations. Unfortunately, the weak decrease of the Airy function for the negative values of the variable (as $x^{-1/4}$ in average) deters the use of this transformation as a numerical filter.

However, as we have seen in the preceding paragraph, an integral representation of the product $Ai(x)Ai(-x)$ corresponds to the Fourier transform

$$Ai(x)Ai(-x) = \frac{1}{2^{4/3}\pi} \int_{-\infty}^{+\infty} Ai\left[2^{-4/3}u^2\right] \mathrm{e}^{\mathrm{i}xu} \mathrm{d}u. \tag{2.166}$$

This product and its Fourier transform are well localized with a fast decrease, displaying the behaviour

$$2^{-4/3} Ai\left[2^{-4/3}u^2\right] \to \frac{1}{8\sqrt{\pi\,|u|}} e^{-\frac{1}{6}|u|^3}$$

as $u \to \pm\infty$. This product and its Fourier transform then form a couple of square integrable functions.

Since the decrease of the Fourier transform is faster than a Gaussian, we can consider this product as a continuous basis of wavelets [Holschneider (1995)]. As a matter of fact, when we take twice the derivative of $Ai(x)Ai(-x)$, we obtain

$$Ai'(x)Ai'(-x) = \frac{1}{2^{4/3}\pi} \int_{-\infty}^{+\infty} Ai\left[2^{-4/3}u^2\right] \mathrm{e}^{\,\mathrm{i}xu} u^2 \mathrm{d}u. \tag{2.167}$$

The first moments of this function cancel:

$$\int_{-\infty}^{+\infty} Ai'(x)Ai'(-x)\, x^n \,\mathrm{d}x = 0, \qquad n = 0, 1, \tag{2.168}$$

i.e. its Fourier transform behaves parabolically near the origin.

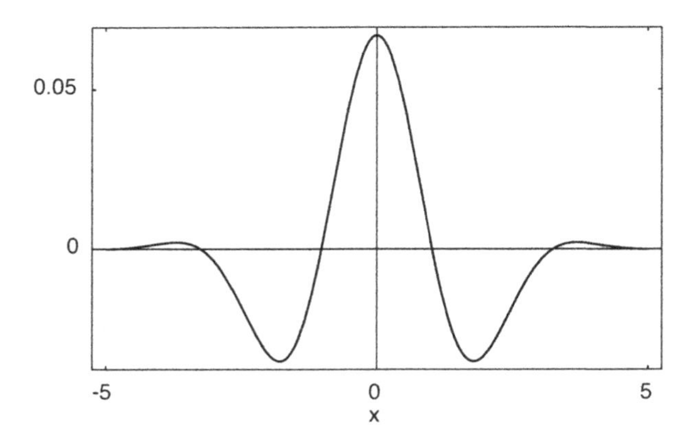

Fig. 2.8 Plot of the wavelets $Ai'(x)Ai'(-x)$.

Figures (2.8) and (2.9) show the wavelet $Ai'(x)Ai'(-x)$ and its Fourier transform (for the positive values of the Fourier variable). The maximum value of the Fourier transform is given by the solution to the equation: $Ai(x_0) + x_0 Ai'(x_0) = 0$, i.e. $x_0 = 0.88405$, and $u_0 = 1.4925$ for the Fourier variable.

This continuous basis of wavelets is very close to what is called the "Mexican hat", or Maar wavelets, but differs in a faster decrease of the Fourier transform for the Airy wavelets.

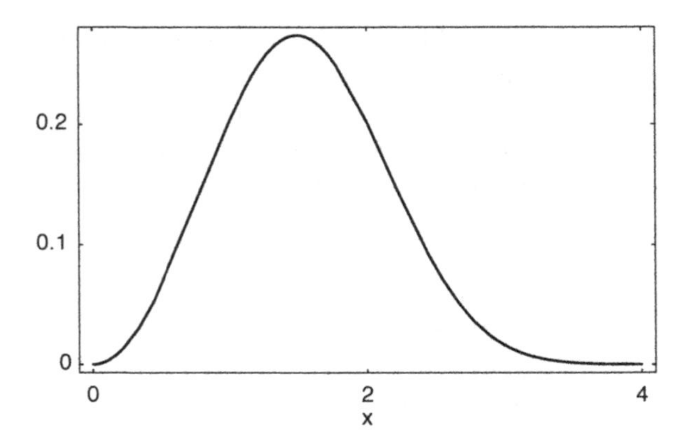

Fig. 2.9 Fourier transform of the wavelets $Ai'(x)Ai'(-x)$.

Finally, we give the normalisation (see Eq. (2.166))

$$\frac{1}{\alpha}\int_{-\infty}^{+\infty} Ai^2\left(\frac{x-a}{\alpha}\right)Ai^2\left(\frac{a-x}{\alpha}\right)\mathrm{d}x = \frac{1}{12\pi}, \tag{2.169}$$

whose demonstration will be given in §3.6.4 (see also in Reid (1995)).

Exercises

(1) Prove, with an appropriate change of function on the Airy differential equation (Eq. (2.1)), that the logarithmic derivative of $Ai(x)$ satisfies the Riccati equation $u' + u^2 = x$. Find a differential equation of the first order for which the solution is $\frac{Ai(x)}{Ai'(x)}$. Conclude in relation with §2.2.2.

(2) Find the solution to the differential equation

$$u\frac{\mathrm{d}u}{\mathrm{d}x} = 2xu + 1$$

in terms of Airy functions (choose a convenient boundary condition). *Hint*: see Davis (1962) in connection with the period of the van der Pol oscillator.

(3) Show that if $\xi = \frac{2}{3}x^{3/2}$

$$Ai(x) = \frac{1}{2\sqrt{\pi}}x^{-1/4}W_{0,1/3}(2\xi),$$

and

$$Bi(x) = \frac{1}{2^{1/3}} \left(-\frac{2}{3}\right) x^{-1/4} M_{0,-1/3}(2\xi),$$

where $W_{\lambda,\mu}(.)$ and $M_{\lambda,\mu}(.)$ are the Whittaker functions. Find similar relations for the derivatives of Airy functions. Find relations between Airy functions and the confluent hypergeometric functions. *Hint*: see the NIST Digital Library of Mathematical Functions — Chap. 9 Airy & Related Functions by F. W. J. Olver. Internet address: http://dlmf.nist.gov/

(4) Find a particular solution to the differential equation

$$y'' - xy = Ai(x).$$

Hint: Calculate the derivatives of $Ai(x)$ up to the third order.

(5) From the Wronskian relation

$$W\{Ai(x),\ Bi(x)\} = \frac{1}{\pi},$$

prove that the Wronskian of squares and product of Airy functions is

$$W\left\{Ai^2(x), Ai(x)Bi(x), Bi^2(x)\right\} = \frac{2}{\pi^3},$$

see Eq. (2.149). *Hint*: The Wronskian of a canonical differential equation of the third order is given in §6.2.1.

Chapter 3

Primitives and Integrals of Airy Functions

3.1 Primitives containing one Airy function

3.1.1 *In terms of Airy functions*

From the formulae (2.129) and (2.130), we deduce the expressions of the primitives of $Ai(x)$ and $Bi(x)$ [Abramowitz & Stegun (1965)]

$$\begin{aligned}\int_0^x Ai(t)\mathrm{d}t &= \frac{1}{3} + \pi\left[Ai'(x)\,Gi(x) - Ai(x)\,Gi'(x)\right] \\ &= -\frac{2}{3} - \pi\left[Ai'(x)\,Hi(x) - Ai(x)\,Hi'(x)\right]\end{aligned} \tag{3.1}$$

$$\begin{aligned}\int_0^x Ai(-t)\mathrm{d}t &= -\frac{1}{3} - \pi\left[Ai'(-x)\,Gi(-x) - Ai(-x)\,Gi'(-x)\right] \\ &= \frac{2}{3} + \pi\left[Ai'(-x)\,Hi(-x) - Ai(-x)\,Hi'(-x)\right]\end{aligned} \tag{3.2}$$

$$\begin{aligned}\int_0^x Bi(t)\mathrm{d}t &= \pi\left[Bi'(x)\,Gi(x) - Bi(x)\,Gi'(x)\right] \\ &= -\pi\left[Bi'(x)\,Hi(x) - Bi(x)\,Hi'(x)\right]\end{aligned} \tag{3.3}$$

$$\begin{aligned}\int_0^x Bi(-t)\mathrm{d}t &= -\pi\left[Bi'(-x)\,Gi(-x) - Bi(-x)\,Gi'(-x)\right] \\ &= \pi\left[Bi'(-x)\,Hi(-x) - Bi(-x)\,Hi'(-x)\right].\end{aligned} \tag{3.4}$$

3.1.2 *Ascending series*

The ascending series of the primitives of Airy functions are [Abramowitz & Stegun (1965)]

$$\int_0^x Ai\,(t)\mathrm{d}t = c_1 F\,(x) - c_2 G\,(x) \tag{3.5}$$

$$\int_0^x Ai\,(-t)\mathrm{d}t = -c_1 F\,(-x) + c_2 G\,(-x) \tag{3.6}$$

$$\int_0^x Bi\,(t)\mathrm{d}t = \sqrt{3}\,[c_1 F\,(x) + c_2 G\,(x)] \tag{3.7}$$

$$\int_0^x Bi\,(-t)\mathrm{d}t = -\sqrt{3}\,[c_1 F\,(-x) + c_2 G\,(-x)]\,, \tag{3.8}$$

with the series $F(x)$ and $G(x)$ being defined by term–by–term integration of the series f and g (cf. §2.1.4)

$$\begin{aligned} F\,(x) &= \sum_{k=0}^{\infty} 3^k \left(\frac{1}{3}\right)_k \frac{x^{3k+1}}{(3k+1)!} \\ &= x + \frac{1}{4!}x^4 + \frac{1.4}{7!}x^7 + \frac{1.4.7}{10!}x^{10} + \ldots \end{aligned}$$

$$\begin{aligned} G\,(x) &= \sum_{k=0}^{\infty} 3^k \left(\frac{2}{3}\right)_k \frac{x^{3k+2}}{(3k+2)!} \\ &= \frac{1}{2!}x^2 + \frac{2}{5!}x^5 + \frac{2.5}{8!}x^8 + \frac{2.5.8}{11!}x^{11} + \ldots \end{aligned}$$

where the constants c_1 and c_2 are defined in §2.1.4.2; $c_1 = Ai(0)$ and $c_2 = Ai'(0)$.

3.1.3 *Asymptotic expansions*

For $x \gg 1$ (and $\xi = \frac{2}{3}x^{3/2}$), the first terms of the asymptotic expansions of the primitives of the homogeneous Airy functions are [Abramowitz & Stegun (1965)]

$$\int_0^x Ai(t)\mathrm{d}t \approx \frac{1}{3} - \frac{\mathrm{e}^{-\xi}}{2\sqrt{\pi}x^{3/4}} \left(1 - \frac{41}{48x^{3/2}} + \frac{9\,241}{4\,608x^3} - \ldots\right) \tag{3.9}$$

$$\int_0^x Ai(-t)\mathrm{d}t \approx \frac{2}{3} + \frac{1}{\sqrt{\pi}x^{3/4}}\left[\left(\frac{7}{48x^{3/2}} - \frac{5}{48x^3} + \dots\right)\sin\left(\xi - \frac{\pi}{4}\right)\right. \quad (3.10)$$
$$\left. - \left(1 + \frac{1}{x^{3/2}} - \frac{8\,761}{4\,608x^3} - \dots\right)\cos\left(\xi - \frac{\pi}{4}\right)\right]$$

$$\int_0^x Bi(t)\mathrm{d}t \approx \frac{\mathrm{e}^{\xi}}{\sqrt{\pi}x^{3/4}}\left(1 + \frac{41}{48x^{3/2}} + \frac{9\,241}{4\,608x^3}\dots\right) \quad (3.11)$$

$$\int_0^x Bi(-t)\mathrm{d}t \approx \frac{1}{\sqrt{\pi}x^{3/4}}\left[\left(\frac{7}{48x^{3/2}} - \frac{5}{48x^3} + \dots\right)\cos\left(\xi - \frac{\pi}{4}\right)\right. \quad (3.12)$$
$$\left. + \left(1 + \frac{1}{x^{3/2}} - \frac{8\,761}{4\,608x^3} + \dots\right)\sin\left(\xi - \frac{\pi}{4}\right)\right].$$

These expansions are obtained by integrating, term–by–term, the expansions defined in §2.1.4.

3.1.4 *Primitives of Scorer functions*

Gordon (1970) also gives some primitives implying the inhomogeneous function $Gi(x)$. The primitive $\int Gi\left[\alpha(x+\beta)\right]\mathrm{d}x$ seems unable to be expressed simply in terms of Airy functions. Nevertheless, we can calculate

- $$\int xGi\left[\alpha(x+\beta)\right]\mathrm{d}x = \frac{x}{\alpha\pi} + \frac{1}{\alpha^2}Gi'\left[\alpha(x+\beta)\right] - \beta\int Gi\left[\alpha(x+\beta)\right]\mathrm{d}x \quad (3.13)$$

- $$\int x^2Gi\left[\alpha(x+\beta)\right]\mathrm{d}x \quad (3.14)$$
$$= \frac{x}{\alpha\pi}\left(\frac{x}{2} - \beta\right) + \frac{x-\beta}{\alpha^2}Gi'\left[\alpha(x+\beta)\right] - \frac{1}{\alpha^3}Gi\left[\alpha(x+\beta)\right] + \beta^2\int Gi\left[\alpha(x+\beta)\right]\mathrm{d}x$$

- $$\int x^3Gi\left[\alpha(x+\beta)\right]\mathrm{d}x = \frac{x}{\alpha\pi}\left(\frac{x^2}{3} - \beta\frac{x}{2} + \beta^2\right) \quad (3.15)$$
$$+ \frac{x^2 - \beta x + \beta^2}{\alpha^2}Gi'\left[\alpha(x+\beta)\right] + \frac{\beta - 2x}{\alpha^3}Gi\left[\alpha(x+\beta)\right] + \left(2\alpha^{-3} - \beta^3\right)\int Gi\left[\alpha(x+\beta)\right]\mathrm{d}x.$$

3.1.5 *Repeated primitives*

For all the primitives given below, the integration constant has been omitted. If y is any linear combination of Airy function and y' its derivative, we denote its primitive by y_1. We then have

- $$\int y_1 \mathrm{d}x = x\, y_1 - y' \tag{3.16}$$
- $$\int x\, y_1 \mathrm{d}x = \frac{1}{2}(x^2 y_1 - x\, y' + y) \tag{3.17}$$
- $$\int x^2\, y_1 \mathrm{d}x = \frac{1}{3}(x^3 y_1 - x\, y' + 2(x\, y - y_1)). \tag{3.18}$$

From which we find

$$\int^x \int^{x'} y_1 \mathrm{d}x' \mathrm{d}x'' = \frac{1}{2}(x^2 y_1 - x\, y' - y), \quad etc. \tag{3.19}$$

3.2 Product of Airy functions

It is sometimes possible to easily calculate the primitive of a product of Airy functions. For example, we can calculate

$$I = \int_x^\infty Ai^2(x)\mathrm{d}x,$$

from an integration by parts

$$I = \left[xAi^2(x)\right]_x^\infty - 2\int_x^\infty xAi(x)Ai'(x)\mathrm{d}x.$$

Thanks to the Airy equation (2.1), we can write

$$\begin{aligned} I &= \left[xAi^2(x)\right]_x^\infty - 2\int_x^\infty Ai'(x)Ai''(x)\mathrm{d}x \\ &= \left[xAi^2(x)\right]_x^\infty - \left[Ai'^2(x)\right]_x^\infty . \end{aligned}$$

Since $\lim_{x\to\infty} xAi^2(x) = 0$ and $\lim_{x\to\infty} Ai'(x) = 0$, we finally obtain

$$I = -xAi^2(x) + Ai'^2(x).$$

However, this kind of calculation is not always so straightforward. We shall therefore detail the method of Albright (1977) in the next section. This method allows us to calculate the primitives of linear combinations of the homogeneous Airy functions Ai and Bi.

3.2.1 *The method of Albright*

We wish to calculate integrals of the kinds

$$\int x^n y^2 \mathrm{d}x, \quad \int x^n y' y \, \mathrm{d}x \text{ and } \int x^n y'^2 \mathrm{d}x, \tag{3.20}$$

where y is a linear combination of the functions $Ai(x)$ and $Bi(x)$ (i.e. $y(x) = \alpha Ai(x) + \beta Bi(x)$), the sign " *prime* " ($'$) stands for differentiation with respect to x.

Albright builds the following table, where D stands for the operator $\mathrm{d}/\mathrm{d}x$, and where y'' is replaced by xy (according to the Airy equation (2.1))

	y^2	$y'y$	y'^2	xy^2	$xy'y$	xy'^2	x^2y^2	$x^2y'y$	$x^2y'^2$
$\mathrm{D}y^2$		2							
$\mathrm{D}y'y$			1	1					
$\mathrm{D}y'^2$					2				
$\mathrm{D}xy^2$	1				2				
$\mathrm{D}xy'y$		1				1	1		
$\mathrm{D}xy'^2$			1					2	
$\mathrm{D}x^2y^2$				2				2	
$\mathrm{D}x^2y'y$					2				1
$\mathrm{D}x^2y'^2$						2			

The properties of this table are such that we are able to calculate the primitives of the kinds in (3.20). For example, to calculate

$$F = \int y^2 \mathrm{d}x,$$

we just have to subtract lines (4) and (3) of the table. So we obtain

$$\mathrm{D}\left(xy^2 - y'^2\right) = y^2,$$

and then the result, Eq. (3.24)

$$F = xy^2 - y'^2,$$

except for the integration constant. For the following primitive

$$G = \int y'^2 \mathrm{d}x,$$

three lines of the table have to be considered; these can be written into the matrix form

$$D \begin{pmatrix} y'y \\ xy'^2 \\ x^2y^2 \end{pmatrix} = \begin{pmatrix} 1 & 1 & 0 \\ 1 & 0 & 2 \\ 0 & 2 & 2 \end{pmatrix} \begin{pmatrix} y'^2 \\ xy^2 \\ x^2y'y \end{pmatrix}. \tag{3.21}$$

It is then sufficient to inverse the system

$$\begin{pmatrix} y'^2 \\ xy^2 \\ x^2y'y \end{pmatrix} = \frac{1}{6}\begin{pmatrix} -4 & -2 & 2 \\ -2 & 2 & -2 \\ 2 & -2 & -1 \end{pmatrix} D \begin{pmatrix} y'y \\ xy'^2 \\ x^2y^2 \end{pmatrix} \tag{3.22}$$

leading immediately to the result

$$G = \frac{1}{3}\left(2y'y + xy'^2 - x^2y^2\right).$$

In the next section we set out some particular cases of primitives obtained by the method of Albright and with the help of the primitive

$$L = \int x^n y'y \mathrm{d}x,$$

calculated from an integration by parts

$$L = x^n y^2 - \int n x^{n-1} y^2 \mathrm{d}x - L,$$

$$L = \frac{1}{2}\left(x^n y^2 - n\int x^{n-1}y^2 \mathrm{d}x\right). \tag{3.23}$$

3.2.2 *Some primitives*

If y is a linear combination of $Ai(x)$ and $Bi(x)$ and n a positive integer:

- $$\int y^2 \mathrm{d}x = xy^2 - y'^2 \tag{3.24}$$
- $$\int xy^2 \mathrm{d}x = \frac{1}{3}\left(y'y - xy'^2 + x^2y^2\right) \tag{3.25}$$
- $$\int x^2y^2 \mathrm{d}x = \frac{1}{5}\left[2\left(xy'y - \frac{1}{2}y^2\right) - x^2y'^2 + x^3y^2\right] \tag{3.26}$$
  - $$\int y'y \mathrm{d}x = \frac{1}{2}y^2 \tag{3.27}$$
  - $$\int xy'y \mathrm{d}x = \frac{1}{2}y'^2 \tag{3.28}$$
  - $$\int x^2y'y \mathrm{d}x = \frac{1}{6}\left(x^2y^2 - 2y'y + 2xy'^2\right) \tag{3.29}$$
- $$\int y'^2 \mathrm{d}x = \frac{1}{3}\left(2y'y + xy'^2 - x^2y^2\right) \tag{3.30}$$
- $$\int xy'^2 \mathrm{d}x = \frac{1}{5}\left[3\left(xy'y - \frac{1}{2}y^2\right) + x^2y'^2 - x^3y^2\right] \tag{3.31}$$
- $$\int x^2y'^2 \mathrm{d}x = \frac{1}{7}\left(4x^2y'y - 4y'^2 + x^3y'^2 - x^4y^2\right) \tag{3.32}$$

$$\bullet \int x^n y^2 \mathrm{d}x = \frac{1}{2n+1}\left(nx^{n-1}y'y - n(n-1)\int x^{n-2}y'y\mathrm{d}x - x^n y'^2 + x^{n+1}y^2\right) \tag{3.33}$$

$$\bullet \int x^n y'y\mathrm{d}x = \frac{1}{2}\left(x^n y^2 - n\int x^{n-1}y^2\mathrm{d}x\right) \tag{3.34}$$

or

$$\bullet \int x^n y'y\mathrm{d}x = \frac{1}{2n-1}\left[\frac{1}{2}nx^{n-1}y'^2 + \frac{1}{2}(n-1)x^n y^2 - \frac{1}{2}n(n-1)\left(x^{n-2}y'y - (n-2)\int x^{n-3}y'y\mathrm{d}x\right)\right] \tag{3.35}$$

$$\bullet \int x^n y'^2\mathrm{d}x = \frac{1}{2n+3}\left[x^{n+1}y'^2 - x^{n+2}y^2 + (n+2)\left(x^n y'y - n\int x^{n-1}y'y\mathrm{d}x\right)\right]$$

If $A(x)$ and $B(x)$ are any two linear combinations of $Ai(x)$ and $Bi(x)$, and n is a positive integer

$$\bullet \int A(x)B(x)\mathrm{d}x = xA(x)B(x) - A'(x)B'(x) \tag{3.36}$$

$$\bullet \int A'(x)B(x)\mathrm{d}x = \frac{1}{2}\left[A(x)B(x) + xA'(x)B(x) - xA(x)B'(x)\right] \tag{3.37}$$

$$\bullet \int A'(x)B'(x)\mathrm{d}x = \frac{1}{3}\left[A'(x)B(x) + A(x)B'(x) + xA'(x)B'(x) - x^2A(x)B(x)\right] \tag{3.38}$$

$$\bullet \int xA(x)B(x)\mathrm{d}x = \frac{1}{6}\left[A'(x)B(x) + A(x)B'(x) - 2xA'(x)B'(x) + 2x^2A(x)B(x)\right] \tag{3.39}$$

(3.40)

- $$\int xA'(x)B(x)\mathrm{d}x = \frac{1}{4}\left[2A'(x)B'(x) + x^2A'(x)B(x) - x^2A(x)B'(x)\right] \tag{3.41}$$

- $$\int xA'(x)B'(x)\mathrm{d}x = \frac{1}{5}\left[\frac{3}{2}\Big(xA'(x)B(x) + xA(x)B'(x) - A(x)B(x)\Big) + x^2A'(x)B'(x) - x^3A(x)B(x)\right] \tag{3.42}$$

- $$\int x^nA(x)B(x)\mathrm{d}x \tag{3.43}$$
$$= \frac{1}{2(2n+1)}\Big[nx^{n-1}\left(A'(x)B(x) + A(x)B'(x)\right) - 2x^nA'(x)B'(x) + 2x^{n+1}A(x)B(x) - n(n-1)\int x^{n-2}\left(A'(x)B(x) + A(x)B'(x)\right)\mathrm{d}x\Big]$$

- $$\int x^nA(x)B(x)\mathrm{d}x \tag{3.44}$$
$$= \frac{1}{2}\left[x^{n-1}\left(A'(x)B(x) + A(x)B'(x)\right) - 2\int x^{n-1}A'(x)B'(x)\mathrm{d}x - (n-1)\int x^{n-2}\left(A'(x)B(x) + A(x)B'(x)\right)\mathrm{d}x\right], \; n \geq 2$$

- $$\int x^nA'(x)B(x)\mathrm{d}x = \frac{1}{2}\left[x^nA(x)B(x) - n\int x^{n-1}A(x)B(x)\mathrm{d}x + \frac{x^{n+1}}{n+1}\left(A'(x)B(x) - A(x)B'(x)\right)\right] \tag{3.45}$$

- $$\int x^nA'(x)B'(x)\mathrm{d}x = \frac{1}{2(2n+3)}\Big\{2x^{n+1}\Big(A'(x)B'(x) - xA(x)B(x)\Big) + (n+2)\Big[x^n\left(A'(x)B(x) + A(x)B'(x)\right) - n\int x^{n-1}\left(A'(x)B(x) + A(x)B'(x)\right)\mathrm{d}x\Big]\Big\}. \tag{3.46}$$

The formulae (3.36) to (3.46) are true for any $A(x)$ and $B(x)$. In particular, if $A(x) = Ai(x)$ and $B(x) = Bi(x)$, we can simplify some of these expressions by using the Wronskian relationship (formula (2.6)),

$$AiBi' - Ai'Bi = \frac{1}{\pi}.$$

For example, the formula (3.37) becomes

$$\int Ai'Bi\,\mathrm{d}x = \frac{1}{2}\left(AiBi - \frac{x}{\pi}\right).$$

Some of the preceding primitives can be calculated for a more general form [Gordon (1969), (1970), (1971)]. If A is a linear combination of $Ai(x)$ and $Bi(x)$, we have the following primitives

- $$\int A\left[\alpha(x+\beta)\right]\mathrm{d}x = \frac{\pi}{\alpha}\left\{A'\left[\alpha(x+\beta)\right]Gi\left[\alpha(x+\beta)\right] - A\left[\alpha(x+\beta)\right]Gi'\left[\alpha(x+\beta)\right]\right\} \tag{3.47}$$

- $$\int xA\left[\alpha(x+\beta)\right]\mathrm{d}x = \frac{1}{\alpha^2}A'\left[\alpha(x+\beta)\right] - \beta\int A\left[\alpha(x+\beta)\right]\mathrm{d}x \tag{3.48}$$

- $$\int x^2A\left[\alpha(x+\beta)\right]\mathrm{d}x = \frac{x-\beta}{\alpha^2}A'\left[\alpha(x+\beta)\right] - \frac{1}{\alpha^3}A\left[\alpha(x+\beta)\right] + \beta^2\int A\left[\alpha(x+\beta)\right]\mathrm{d}x \tag{3.49}$$

- $$\int x^3A\left[\alpha(x+\beta)\right]\mathrm{d}x = \frac{x^2-\beta x+\beta^2}{\alpha^2}A'\left[\alpha(x+\beta)\right] - \frac{\beta-2x}{\alpha^3}A\left[\alpha(x+\beta)\right] + \left(2\alpha^{-3}-\beta^3\right)\int A\left[\alpha(x+\beta)\right]\mathrm{d}x \tag{3.50}$$

When A and B are linear combinations of $Ai(x)$ and $Bi(x)$, we have the primitives

- $$\int A\left[\alpha(x+\beta)\right]B\left[\alpha(x+\beta)\right]\mathrm{d}x = (x+\beta)A\left[\alpha(x+\beta)\right]B\left[\alpha(x+\beta)\right] - \frac{1}{\alpha}A'\left[\alpha(x+\beta)\right]B'\left[\alpha(x+\beta)\right] \tag{3.51}$$

$$\bullet \int xA\left[\alpha(x+\beta)\right]B\left[\alpha(x+\beta)\right]\mathrm{d}x \tag{3.52}$$

$$\begin{aligned}
&= \frac{1}{3}\left(x^2 - x\beta - 2\beta^2\right)A\left[\alpha(x+\beta)\right]B\left[\alpha(x+\beta)\right] \\
&\quad + \frac{1}{6\alpha^2}\left\{A'\left[\alpha(x+\beta)\right]B\left[\alpha(x+\beta)\right]\right. \\
&\qquad \left. + A\left[\alpha(x+\beta)\right]B'\left[\alpha(x+\beta)\right]\right\} \\
&\quad + \frac{2\beta - x}{3\alpha}A'\left[\alpha(x+\beta)\right]B'\left[\alpha(x+\beta)\right]
\end{aligned}$$

$$\bullet \int x^2 A\left[\alpha(x+\beta)\right]B\left[\alpha(x+\beta)\right]\mathrm{d}x \tag{3.53}$$

$$\begin{aligned}
&= \frac{1}{15}\left(3x^3 - x^2\beta + 4x\beta^2 + 8\beta^3 - 3\alpha^{-3}\right) \\
&\qquad \times\ A\left[\alpha(x+\beta)\right]B\left[\alpha(x+\beta)\right] \\
&\quad + \frac{3x - 2\beta}{15\alpha^2}\left\{A'\left[\alpha(x+\beta)\right]B\left[\alpha(x+\beta)\right]\right. \\
&\qquad \left. + A\left[\alpha(x+\beta)\right]B'\left[\alpha(x+\beta)\right]\right\} \\
&\quad - \frac{3x^2 - 4x\beta + 8\beta^2}{15\alpha}A'\left[\alpha(x+\beta)\right]B'\left[\alpha(x+\beta)\right]
\end{aligned}$$

$$\bullet \int A\left[\alpha(x+\beta_1)\right]B\left[\alpha(x+\beta_2)\right]\mathrm{d}x \tag{3.54}$$

$$\begin{aligned}
&= \frac{1}{\alpha^2\left(\beta_1 - \beta_2\right)}\left\{A'\left[\alpha(x+\beta_1)\right]B\left[\alpha(x+\beta_2)\right]\right. \\
&\qquad \left. - A\left[\alpha(x+\beta_1)\right]B'\left[\alpha(x+\beta_2)\right]\right\}
\end{aligned}$$

$$\bullet \int xA\left[\alpha(x+\beta_1)\right]B\left[\alpha(x+\beta_2)\right]\mathrm{d}x \tag{3.55}$$

$$\begin{aligned}
&= -\frac{\beta_1 + \beta_2 + 2x}{\alpha^3\left(\beta_1 - \beta_2\right)^2}A\left[\alpha(x+\beta_1)\right]B\left[\alpha(x+\beta_2)\right] \\
&\quad + \left[\frac{x}{\alpha^2\left(\beta_1 - \beta_2\right)} + \frac{2}{\alpha^5\left(\beta_1 - \beta_2\right)^3}\right] \\
&\qquad \times\ \left\{A'\left[\alpha(x+\beta_1)\right]B\left[\alpha(x+\beta_2)\right]\right. \\
&\qquad \left. - A\left[\alpha(x+\beta_1)\right]B'\left[\alpha(x+\beta_2)\right]\right\} \\
&\quad + \frac{2}{\alpha^4\left(\beta_1 - \beta_2\right)^2}A'\left[\alpha(x+\beta_1)\right]B'\left[\alpha(x+\beta_2)\right].
\end{aligned}$$

For completeness, we also give the integral[1]

$$\bullet \int x^2\, A[\alpha(x+\beta_1)]\, B[\alpha(x+\beta_2)]\, \mathrm{d}x \tag{3.56}$$

$$\begin{aligned}
&= \frac{4}{(\beta_2-\beta_1)^2}\left\{-\frac{1}{\alpha^3}\left[x^2+c\,x+\frac{3(\beta_1+\beta_2)}{\alpha^3\,(\beta_2-\beta_1)^2}\right]A[\alpha(x+\beta_1)]B[\alpha(x+\beta_2)]\right.\\
&\quad-\left[\frac{(\beta_2-\beta_1)\,x^2}{4\,\alpha^2}+\frac{3\,x+\beta_2+c}{\alpha^5\,(\beta_2-\beta_1)}\right]A'[\alpha(x+\beta_1)]\,B[\alpha(x+\beta_2)]\\
&\quad+\left[\frac{(\beta_2-\beta_1)\,x^2}{4\,\alpha^2}+\frac{3\,x+\beta_1+c}{\alpha^5\,(\beta_2-\beta_1)}\right]A[\alpha(x+\beta_1)]\,B'[\alpha(x+\beta_2)]\\
&\quad\left.+\frac{1}{\alpha^4}\left[x+\frac{6}{\alpha^3\,(\beta_2-\beta_1)^2}\right]A'[\alpha(x+\beta_1)]\,B'[\alpha(x+\beta_2)]\right\},
\end{aligned}$$

with

$$c=\frac{(\beta_2-\beta_1)^2\,(\beta_2+\beta_1)+12/\alpha^3}{2\,(\beta_2-\beta_1)^2}.$$

Finally, we give some primitives involving the primitive y_1 of any linear combination of Airy functions y,

$$\bullet \int y_1^2\,\mathrm{d}x = x\,y_1^2 - 2y'y_1 + y^2 \tag{3.57}$$

$$\bullet \int y_1 y\,\mathrm{d}x = \frac{1}{2}\,y_1^2 \tag{3.58}$$

$$\bullet \int y_1 y'\,\mathrm{d}x = y_1 y - x\,y^2 + y'^2, \tag{3.59}$$

where y' is the derivative of y with respect to x.

3.3 Other primitives

Albright & Gavathas (1986) give other kinds of primitives of Airy functions. The expression

$$\int \frac{Ai(x)Bi(x)}{\left[Ai^2(x)+Bi^2(x)\right]^2}\mathrm{d}x = \frac{\pi}{2}\,\frac{Bi^2(x)}{Ai^2(x)+Bi^2(x)}, \tag{3.60}$$

is obtained by differentiation and thanks to the Wronskian of Ai and Bi (formula (2.6)). More generally, we have

$$\int \frac{Ai^{n-1}(x)Bi^{n-1}(x)}{\left[Ai^n(x)+Bi^n(x)\right]^2}\mathrm{d}x = \frac{\pi}{n}\,\frac{Bi^n(x)}{Ai^n(x)+Bi^n(x)}, \tag{3.61}$$

[1]It is not given in the work of Gordon.

and, from a similar method, we obtain

$$\int \frac{\mathrm{d}x}{Ai^2(x)} = \pi \frac{Bi(x)}{Ai(x)} \tag{3.62}$$

$$\int \frac{\mathrm{d}x}{Bi^2(x)} = -\pi \frac{Ai(x)}{Bi(x)} \tag{3.63}$$

$$\int \frac{\mathrm{d}x}{Ai(x)Bi(x)} = \pi \ln \frac{Bi(x)}{Ai(x)} \tag{3.64}$$

$$\int \frac{Bi^n(x)}{Ai^{n+2}(x)} \mathrm{d}x = \frac{\pi}{n+1} \left(\frac{Bi(x)}{Ai(x)} \right)^{n+1} . \tag{3.65}$$

Albright & Gavathas (1986) build more general results by considering two functions: namely f and F such that $f = F'$. We have then

$$\int \frac{1}{Ai^2(x)} f \left(\frac{Bi(x)}{Ai(x)} \right) \mathrm{d}x = \pi F \left(\frac{Bi(x)}{Ai(x)} \right), \tag{3.66}$$

and

$$\int \frac{1}{Bi^2(x)} f \left(\frac{Ai(x)}{Bi(x)} \right) \mathrm{d}x = -\pi F \left(\frac{Ai(x)}{Bi(x)} \right). \tag{3.67}$$

These results are easily verified. The equations (3.60) to (3.65) are nothing but specific cases of this general result. There are also some interesting particular cases, such as

$$\int \frac{\mathrm{d}x}{Ai^2(x) + Bi^2(x)} = \pi \tan^{-1} \frac{Bi(x)}{Ai(x)}, \tag{3.68}$$

and

$$\int \frac{Bi^n(x)}{Ai^{n+2}(x)} \exp \left(\frac{Bi(x)}{Ai(x)} \right)^{n+1} \mathrm{d}x = \frac{\pi}{n+1} \exp \left(\frac{Bi(x)}{Ai(x)} \right)^{n+1} . \tag{3.69}$$

We have the general relation

$$F\left(\frac{u}{v}\right) = \int \frac{W[v,u]}{v^2} f\left(\frac{u}{v}\right) \mathrm{d}x, \tag{3.70}$$

where $W[v,u]$ is the Wronskian of v and u, and $F' = f$. From the primitive of the Airy function Ai:

$$Ai_1(x) = \int_x^\infty Ai(z)\, \mathrm{d}z,$$

Varlamov (2008-b) gives several indefinite integrals involving the Scorer function Gi. First we have the Wronskian $W[Ai(x), Gi(x)] = \frac{1}{\pi} Ai_1(x)$ (see Eq. (3.47)), and then the general formula

$$\int \frac{Ai_1(x)}{Gi^2(x)} f \left(\frac{Ai(x)}{Gi(x)} \right) \mathrm{d}x = -\pi F \left(\frac{Ai(x)}{Gi(x)} \right), \tag{3.71}$$

from which many indefinite integrals may be found, such as

$$\int \frac{Ai_1(x)}{Gi^2(x)}\,\mathrm{d}x = -\pi\frac{Ai(x)}{Gi(x)},$$

$$\int \frac{Ai_1(x)}{Ai(x)Gi(x)}\,\mathrm{d}x = -\pi\ln\left(\frac{Ai(x)}{Gi(x)}\right),$$

$$\int \frac{Ai_1(x)}{Ai^2(x)+Gi^2(x)}\,\mathrm{d}x = -\pi\tan^{-1}\left(\frac{Ai(x)}{Gi(x)}\right),$$

and so forth.

3.4 Miscellaneous

Isolated values for primitives involving Airy functions may be given here as well. For instance, we can mention

$$\int \frac{Ai'^2}{x^2}\,\mathrm{d}x = Ai^2 - \frac{Ai'^2}{x}. \tag{3.72}$$

There are also some results related to a solution of the heat equation

$$u(x,t) = \exp(2t^3/3 - xt)\,Ai(t^2 - x).$$

This function has particular and interesting properties. In the case where $x = 0$, for example we have

$$\int \exp(2\,t^3/3)\,Ai(t^2)\,\mathrm{d}t = \exp(2\,t^3/3)\,[t\,Ai(t^2) - Ai'(t^2)], \tag{3.73}$$

$$\int 2t\,\exp(2\,t^3/3)\,[t\,Ai(t^2) + Ai'(t^2)]\,\mathrm{d}t = \exp(2\,t^3/3)\,Ai(t^2), \tag{3.74}$$

$$\int 2t^2\,\exp(2\,t^3/3)\,[t\,Ai(t^2) + Ai'(t^2)]\,\mathrm{d}t = \exp(2\,t^3/3)\,Ai'(t^2). \tag{3.75}$$

We shall return to this solution in §7.4.

3.5 Elementary integrals

3.5.1 *Particular integrals*

From the asymptotic expansion of §3.1.3, we may deduce the values of the defined integrals

$$\int_{-\infty}^{0} Ai(t)\,\mathrm{d}t = \frac{2}{3}, \qquad \int_{0}^{\infty} Ai(t)\,\mathrm{d}t = \frac{1}{3}, \quad \text{then} \int_{-\infty}^{+\infty} Ai(t)\,\mathrm{d}t = 1 \tag{3.76}$$

and

$$\int_{-\infty}^{0} Bi(t)\,\mathrm{d}t = 0, \qquad \int_{0}^{\infty} Bi(t)\,\mathrm{d}t \approx \infty. \tag{3.77}$$

The Airy function Ai is not square integrable on $\mathbb{R}$ but

$$\int_{0}^{\infty} Ai^2(t)\,\mathrm{d}t = \frac{1}{3^{2/3}\Gamma^2(\frac{1}{3})}. \tag{3.78}$$

For the cube of the Airy function, we have [Reid (1997a)]

$$\int_{-\infty}^{+\infty} Ai^3(t)\,\mathrm{d}t = \frac{1}{4\pi^2}\Gamma^2(\frac{1}{3}), \tag{3.79}$$

and

$$\int_{-\infty}^{+\infty} Ai^2(t)Bi(t)\,\mathrm{d}t = \frac{1}{4\pi^2\sqrt{3}}\Gamma^2(\frac{1}{3}). \tag{3.80}$$

Reid (1997b) gives integrals with the fourth power, for example

$$\int_{0}^{+\infty} Ai^4(t), \mathrm{d}t = \frac{1}{24\pi^2}\ln 3. \tag{3.81}$$

The reader may find other integrals in the papers by Reid (1997a,b).

3.5.2 *Integrals containing a single Airy function*

3.5.2.1 *Integrals involving algebraic functions*

Following the notation of Aspnes (1966), we set

$$Ai_1(x) = \int_{x}^{\infty} Ai(t)\mathrm{d}t,$$

which can be expressed in terms of inhomogeneous functions (formulae (3.1) and (3.76)),

$$Ai_1(x) = \pi\left[Ai(x)Gi'(x) - Ai'(x)Gi(x)\right].$$

We now have for $n > 0$

$$\begin{aligned}\int_0^\infty t^n Ai'(t+x)\mathrm{d}t &= -n\int_0^\infty t^{n-1} Ai(t+x)\mathrm{d}t \\ &= \frac{\mathrm{d}}{\mathrm{d}x}\int_0^\infty t^n Ai(t+x)\mathrm{d}t,\end{aligned} \tag{3.82}$$

and for $n > -1$

$$\begin{aligned}\int_0^\infty t^n Ai_1(t+x)\mathrm{d}t &= \frac{1}{n+1}\int_0^\infty t^{n+1} Ai(t+x)\mathrm{d}t \\ &= \int_x^\infty\int_0^\infty t^n Ai(t+x)\mathrm{d}t\mathrm{d}x,\end{aligned} \tag{3.83}$$

with the particular case

$$\int_0^\infty Ai_1(t+x)\mathrm{d}t = Ai'(x) + x\,Ai_1(x). \tag{3.84}$$

These two kinds of integrals can be treated from the calculation of $\int_0^\infty t^n Ai(t+x)\mathrm{d}t$. Now, from the Airy equation (formula (2.1)), we can deduce

$$\begin{aligned}&\int_0^\infty t^n Ai(t+x)\mathrm{d}t \\ &\quad = \int_0^\infty t^{n-1}\frac{\mathrm{d}^2 Ai(t+x)}{\mathrm{d}x^2}\mathrm{d}t - x\int_0^\infty t^{n-1} Ai(t+x)\mathrm{d}t \\ &\quad = \left[\frac{\mathrm{d}^2}{\mathrm{d}x^2} - x\right]\int_0^\infty t^{n-1} Ai(t+x)\mathrm{d}t.\end{aligned} \tag{3.85}$$

In the last two equations, the exponent of the variable t is reduced by a unity until the iteration leads to $0 \geq n > -1$.

It should be noted that, for $x = 0$, the integral (3.85) can be explicitly written [Olver (1974)]

$$\int_0^\infty t^n Ai(t)\mathrm{d}t = \frac{\Gamma(n+1)}{3^{(n+3)/3}\Gamma\left(\frac{n+3}{3}\right)}. \tag{3.86}$$

We can obtain the moments of the Airy function by calculating [Gislason (1973)]

$$\lim_{\varepsilon\to 0}\int_{-\infty}^{+\infty} Ai(x)x^n \mathrm{e}^{-\varepsilon x^2}\mathrm{d}x,\ n\in\mathbb{N},$$

that is to say

$$\int_{-\infty}^{+\infty} Ai(x)\,x^{3k}\mathrm{d}x = \frac{(3k)!}{3^k\,k!} \tag{3.87}$$

$$\int_{-\infty}^{+\infty} Ai(x)\,x^{3k+1}\mathrm{d}x = \int_{-\infty}^{+\infty} Ai(x)\,x^{3k+2}\mathrm{d}x = 0. \tag{3.88}$$

We can also explicitly write the integrals (3.82) to (3.85), for $n = -1/2$, with the help of the definition (2.21) of $Ai(x)$,[2]

$$\int_0^\infty Ai(x+t)\frac{\mathrm{d}t}{\sqrt{t}} = \frac{1}{\pi^{1/2}}\int_0^\infty \cos\left(\frac{u^3}{3}+xu+\frac{\pi}{4}\right)\frac{\mathrm{d}u}{\sqrt{u}}.$$

We obtain, by comparison with Eq. (2.150):

$$\int_0^\infty Ai(x+t)\frac{\mathrm{d}t}{\sqrt{t}} = 2^{2/3}\pi Ai^2\left(\frac{x}{2^{2/3}}\right). \tag{3.89}$$

Here we retrieve the formula (2.157). In a similar manner, we shall have

$$\int_0^\infty Ai(x-t)\frac{\mathrm{d}t}{\sqrt{t}} = 2^{2/3}\pi Ai\left(\frac{x}{2^{2/3}}\right)Bi\left(\frac{x}{2^{2/3}}\right). \tag{3.90}$$

[2]Some of the following integrals may be found, thanks to the relationship between the Fredholm equations $\int_0^\infty f(x+t)\,\frac{\mathrm{d}t}{\sqrt{t}} = g(x) \iff f(x) = -\frac{1}{\pi}\int_0^\infty g'(x+t)\,\frac{\mathrm{d}t}{\sqrt{t}}$.

Differentiating and integrating Eq. (3.85), we obtain

$$\int_0^{\infty} Ai'(x+t)\frac{\mathrm{d}t}{\sqrt{t}} = 2\pi Ai\left(\frac{x}{2^{2/3}}\right) Ai'\left(\frac{x}{2^{2/3}}\right) \tag{3.91}$$

$$\int_0^{\infty} Ai_1(x+t)\frac{\mathrm{d}t}{\sqrt{t}} \tag{3.92}$$

$$= 2^{4/3}\pi\left\{Ai'^2\left(\frac{x}{2^{2/3}}\right) - \frac{x}{2^{2/3}}Ai^2\left(\frac{x}{2^{2/3}}\right)\right\}.$$

We can also mention the integrals [Aspnes (1967)] that are obtained thanks to the causal relations between the Airy functions (cf. §4.1.1)

$$\int_0^{+\infty} Gi(x+t)\frac{\mathrm{d}t}{\sqrt{t}} = 2^{2/3}\pi Ai\left(\frac{x}{2^{2/3}}\right) Bi\left(\frac{x}{2^{2/3}}\right) \tag{3.93}$$

$$\int_0^{+\infty} Gi(x-t)\frac{\mathrm{d}t}{\sqrt{t}} = -2^{2/3}\pi Ai^2\left(\frac{x}{2^{2/3}}\right). \tag{3.94}$$

It should be noted that we can encounter formula (3.89) in a different form [Berry (1977a)]

$$L = \int_{-y}^{\infty} \frac{Ai(x)}{\sqrt{x+y}}\mathrm{d}x = \int_{-\infty}^{+\infty} Ai\left(u^2 - y\right)\mathrm{d}u. \tag{3.95}$$

Thanks to definition (2.21) of the Airy function Ai, we can express this function as

$$L = \frac{1}{2\pi}\iint_{\mathbb{R}^2} \mathrm{e}^{\mathrm{i}\left[v^3/3+(u^2-y)v\right]}\mathrm{d}u\mathrm{d}v.$$

Making the following change of variables

$$\begin{cases} u = 2^{-2/3}\left(Y - X\right) \\ v = 2^{-2/3}\left(Y + X\right), \end{cases}$$

we obtain

$$L = \frac{1}{2\pi 2^{1/3}}\int_{-\infty}^{+\infty} \mathrm{e}^{\mathrm{i}\left[X^3/3+Y^3/3-2^{-2/3}y(X+Y)\right]}\mathrm{d}X\mathrm{d}Y,$$

in other words

$$L = 2^{2/3}\pi Ai^2\left(\frac{y}{2^{2/3}}\right). \tag{3.96}$$

3.5.2.2 *Integrals involving transcendental functions*

Considering first the integral

$$M = \int_{-\infty}^{+\infty} Ai\left(x^2 + a\right) \mathrm{e}^{ikx} \mathrm{d}x, \tag{3.97}$$

that can be written, again using (2.21) for the Ai function,

$$M = \frac{1}{2\pi} \iint_{\mathbb{R}^2} \mathrm{e}^{\mathrm{i}\left[t^3/3 + t\left(x^2+a\right)\right]} \mathrm{e}^{ikx} \mathrm{d}x \mathrm{d}t.$$

The calculation of this integral is not so easy if one integrates first on the variable x. However, it becomes simpler if we employ the method described in §3.6.5, that is to say if we canonise the cubics $t^3/3 + x^2 t$. When we make the following change of variable

$$\begin{cases} X = 2^{-1/3}(t - x) \\ Y = 2^{-1/3}(t + x), \end{cases}$$

we immediately obtain the result (see formulae (2.153) to (2.156))

$$M = 2^{2/3} \pi Ai\left[2^{-2/3}(a - k)\right] Ai\left[2^{-2/3}(a + k)\right]. \tag{3.98}$$

We can also, thanks to the integral representation (2.21) of $Ai(x)$, obtain the relation [Widder (1979)]

$$\int_{-\infty}^{+\infty} \mathrm{e}^{u\,t} Ai\,(t)\,\mathrm{d}t = \mathrm{e}^{u^3/3}. \tag{3.99}$$

Further, we can mention the following integral

$$\int_{-\infty}^{+\infty} \mathrm{e}^{-t^2/4u} Ai\,(t)\,\mathrm{d}t = 2\sqrt{\pi u}\,\mathrm{e}^{2u^3/3} Ai\left(u^2\right). \tag{3.100}$$

In this last expression, we find, thanks to the asymptotic formula (2.45), the relation (3.76) $\int_{-\infty}^{+\infty} Ai(t)\,\mathrm{d}t = 1$, for $u \to \infty$.

We can now give some other integrals involving one Airy function and transcendentals without demonstration

- $$\int_0^\infty Ai(-x\,t)\,\ln t\,\mathrm{d}t = \frac{2}{9x}\left[\psi\left(\frac{2}{3}\right) + \psi\left(\frac{1}{3}\right) - \ln\left(\frac{x^3}{9}\right)\right], \tag{3.101}$$

where $\psi(.)$ stands for the logarithmic derivative of the gamma function.

- $$\int_0^\infty \mathrm{e}^{-2t^3/3x^3} Ai(t^2)\,\mathrm{d}t = \tag{3.102}$$

$$\frac{x^3}{6\sqrt{1-x^6}}\left[\frac{1}{x}\left(1+\sqrt{1-x^6}\right)^{1/3} - x\left(1+\sqrt{1-x^6}\right)^{-1/3}\right].$$

$$\bullet \int_0^{\infty} \mathrm{e}^{-t^3/3}\, Ai\left(-\frac{x^2}{t^2}\right)\,\mathrm{d}t = \frac{2\pi}{\sqrt{3}}\, Ai(x)Ai(-x). \tag{3.103}$$

$$\bullet \int_0^{\infty} \mathrm{e}^{-t^3/12}\, Ai'\,(xt)\,\frac{\mathrm{d}t}{\sqrt{t}} = -\sqrt{\frac{4\pi}{3}}\,\mathrm{e}^{2x^3/3}\, Ai(x^2). \tag{3.104}$$

$$\bullet \int_0^{\infty} \mathrm{e}^{-t^3/12}\, Ai\,(xt)\,\frac{\mathrm{d}t}{t\sqrt{t}} = -\sqrt{\frac{4\pi}{3}}\,\mathrm{e}^{2x^3/3}\,\left(x\,Ai(x^2) - Ai'(x^2)\right). \tag{3.105}$$

$$\bullet \int_0^{\infty} \mathrm{e}^{-t^3/12}\, Ai\,(xt)\; t\sqrt{t}\,\mathrm{d}t = -2\sqrt{\frac{4\pi}{3}}\,\mathrm{e}^{2x^3/3}\,\left(x\,Ai(x^2) + Ai'(x^2)\right). \tag{3.106}$$

3.5.3 *Integrals of products of two Airy functions*

In a similar manner to that in §3.5.2, following the method of Aspnes, we can establish the relation [Aspnes (1966)]

$$\int_0^{\infty} t^n\, Ai^2\,(t+x)\,\mathrm{d}t \tag{3.107}$$

$$= \frac{n}{2n+1}\left[\frac{1}{2}\frac{\mathrm{d}^2}{\mathrm{d}x^2} - 2x\right]\int_0^{\infty} t^{n-1} Ai^2\,(t+x)\,\mathrm{d}t,\ n > 0.$$

We can extend this relation to one containing Ai', since we have $Ai(x)Ai'(x) = \frac{1}{2}\frac{\mathrm{d}}{\mathrm{d}x}Ai^2(x)$ and $Ai'^2(x) = \frac{1}{2}\left[\frac{\mathrm{d}^2}{\mathrm{d}x^2} - x\right]Ai^2(x)$.

In particular, for $n = -1/2$, we find

$$\int_0^{\infty} Ai^2(t+x)\frac{\mathrm{d}t}{\sqrt{t}} = \frac{1}{2}Ai_1\left(2^{2/3}x\right) \tag{3.108}$$

$$\int_0^{\infty} Ai(t+x)Ai'(t+x)\frac{\mathrm{d}t}{\sqrt{t}} = -2^{-4/3}\, Ai\left(2^{2/3}x\right) \tag{3.109}$$

$$\int_0^{\infty} Ai'^2(t+x)\frac{\mathrm{d}t}{\sqrt{t}} \tag{3.110}$$

$$= -\frac{2^{-2/3}}{4}\left\{3Ai'\left(2^{2/3}x\right) + 2^{2/3}x Ai_1\left(2^{2/3}x\right)\right\}.$$

- We can also mention the important integrals

$$\frac{1}{|\alpha\beta|}\int_{-\infty}^{+\infty} Ai\left[\frac{x+a}{\alpha}\right] Ai\left[\frac{x+b}{\beta}\right] \mathrm{d}x \tag{3.111}$$

$$= \begin{cases} \delta(b-a) & \text{if } \beta = \alpha \\ \frac{1}{|\beta^3-\alpha^3|^{1/3}}\, Ai\left[\frac{b-a}{(\beta^3-\alpha^3)^{1/3}}\right] & \text{if } \beta > \alpha, \end{cases}$$

and

$$\frac{1}{|\alpha\beta|}\int_{-\infty}^{+\infty} Ai\left[\frac{x+a}{\alpha}\right] Gi\left[\frac{x+b}{\beta}\right] \mathrm{d}x \tag{3.112}$$

$$= \begin{cases} \frac{1}{\pi}\wp\frac{1}{b-a} & \text{if } \beta = \alpha \\ \frac{1}{|\beta^3-\alpha^3|^{1/3}}\, Gi\left[\frac{b-a}{(\beta^3-\alpha^3)^{1/3}}\right] & \text{if } \beta > \alpha, \end{cases}$$

Eq. (3.112) is calculated from the integral representation (4.7) of $Ai + \mathrm{i}Gi$, $\wp$ representing the Cauchy principal value.

- In order to calculate the following integral [Biennieck (1977)], which is a generalisation of formula (3.111):

$$I_n = \int_{-\infty}^{+\infty} x^n Ai\left[\frac{x+a}{\alpha}\right] Ai\left[\frac{x+b}{\beta}\right] \mathrm{d}x,\ \beta > \alpha,\ \beta > 0, \tag{3.113}$$

the general method consists of taking the integral representation of the Airy functions (2.21), namely

$$Ai\left[\frac{x+a}{\alpha}\right] = \frac{|\alpha|}{2\pi}\int_{-\infty}^{+\infty} \mathrm{e}^{\,\mathrm{i}\left[(\alpha t)^3/3+(x+a)t\right]}\mathrm{d}t. \tag{3.114}$$

Then we obtain

$$I_n = \iint_{\mathbb{R}^2} \mathrm{d}t\mathrm{d}t'\ \frac{|\alpha\beta|}{4\pi^2}\mathrm{e}^{\,\mathrm{i}\left[(\alpha t)^3/3+(\beta t')^3/3+at+bt'\right]}\int_{-\infty}^{+\infty} x^n \mathrm{e}^{\,\mathrm{i}x(t+t')}\mathrm{d}x,$$

with

$$\delta^{(n)}(t+t') = \frac{\mathrm{i}^n}{2\pi}\int_{-\infty}^{+\infty} x^n \mathrm{e}^{\,\mathrm{i}x(t+t')}\mathrm{d}x$$

and

$$\iint f(x)g(y)\delta^{(n)}(x+y)\mathrm{d}x\mathrm{d}y = \int f(x)g^{(n)}(-x)\mathrm{d}x,$$

where δ is the Dirac delta function and $\delta^{(n)}$ its nth derivative. We finally find the relation

$$I_n = \frac{|\alpha\beta|}{2\pi}\int_{-\infty}^{+\infty} \mathrm{e}^{\,\mathrm{i}\left[(\alpha t)^3/3+at\right]}\frac{\mathrm{d}^n}{\mathrm{d}t^n}\left\{\mathrm{i}^{-n}\mathrm{e}^{-\mathrm{i}\left[(\beta t)^3/3+bt\right]}\right\}\mathrm{d}t. \tag{3.115}$$

Another method starts from Eq. (3.113) and from the Airy equation (2.1), leading to the recurrence relation

$$I_{n+1} = \beta^3\frac{\mathrm{d}^2 I_n}{\mathrm{d}b^2} - bI_n \tag{3.116}$$

with I_0 being given by the relation (3.111), we can then deduce

$$I_0 = \frac{|\alpha\beta|}{(\beta^3-\alpha^3)^{1/3}}Ai\left[\frac{b-a}{(\beta^3-\alpha^3)^{1/3}}\right] \tag{3.117}$$

$$I_1 = \frac{b\alpha^3 - a\beta^3}{\beta^3-\alpha^3}I_0 \tag{3.118}$$

$$I_2 = \left[\frac{b\alpha^3-a\beta^3}{\beta^3-\alpha^3}\right]^2 I_0 + \frac{2\alpha^3\beta^3}{\beta^3-\alpha^3}I_0' \tag{3.119}$$

$$\vdots$$

• Note also the following integral [Balazs & Zipfel (1973)], which appears, in particular, in the semiclassical calculation of the Wigner function (cf. §8.3)

$$F = \int_{-\infty}^{+\infty} Ai(x+t)Ai(x-t)\mathrm{e}^{\,\mathrm{i}2pt}\mathrm{d}t. \tag{3.120}$$

To calculate this integral, we again use the integral representation of Ai (2.21) to obtain

$$F = \frac{1}{4\pi^2}\int_{-\infty}^{+\infty} \mathrm{e}^{\,\mathrm{i}\left(u^3/3+v^3/3+ux+vx\right)}\mathrm{e}^{\,\mathrm{i}(2p+u-v)t}\mathrm{d}u\mathrm{d}v\mathrm{d}t.$$

Integration on the variable t leads to a Dirac function, making it possible to derive the relation

$$F = \frac{\mathrm{e}^{\,\mathrm{i}\left(8p^3/3+2px\right)}}{2\pi}\int_{-\infty}^{+\infty} \mathrm{e}^{\,\mathrm{i}\left[2u^3/3+2pu^2+(2x+4p^2)u\right]}\mathrm{d}u,$$

and, after the change of variable $\omega^3 = 2u^3$,

$$F = \frac{\mathrm{e}^{\,\mathrm{i}\left(8p^3/3+2px\right)}}{2\pi} 2^{-1/3} \int_{-\infty}^{+\infty} \mathrm{e}^{\,\mathrm{i}\left[\omega^3/3+2^{1/3}p\omega^2+2^{2/3}(x+2p^2)\omega\right]} \mathrm{d}\omega.$$

The integral

$$G = \int_{-\infty}^{+\infty} \mathrm{e}^{\,\mathrm{i}\left(t^3/3+at^2+bt\right)} \mathrm{d}t, \tag{3.121}$$

is simply calculated by putting $u = t + a$, providing

$$G = \mathrm{e}^{\,\mathrm{i}a\left(2a^2/3-b\right)} \int_{-\infty}^{+\infty} \mathrm{e}^{\,\mathrm{i}\left(u^3/3+(b-a^2)u\right)} \mathrm{d}u,$$

in other words (see formula (2.26))

$$G = 2\pi \mathrm{e}^{\,\mathrm{i}a\left(2a^2/3-b\right)} Ai\left(b - a^2\right). \tag{3.122}$$

In our case, after simplification, we obtain

$$F = 2^{-1/3} Ai\left[2^{2/3}\left(x + p^2\right)\right]. \tag{3.123}$$

Conversely, we have

$$\frac{1}{2\pi} \int_{-\infty}^{+\infty} Ai(a + x^2) \mathrm{e}^{ibx}\, \mathrm{d}x = 2^{-1/3} Ai\left(\frac{a+b}{2^{2/3}}\right) Ai\left(\frac{a-b}{2^{2/3}}\right). \tag{3.124}$$

But also a non-trivial result obtained by Berry *et al.* (1979), and independently by Trinkaus & Drepper (1977).

$$\frac{1}{2\pi} \int_{-\infty}^{+\infty} Ai(a - x^2) \mathrm{e}^{ibx}\, \mathrm{d}x = 2^{-1/3} \Re\left\{Ai\left(\frac{a+\mathrm{i}b}{2^{2/3}}\right) Bi\left(\frac{a-\mathrm{i}b}{2^{2/3}}\right)\right\}. \tag{3.125}$$

More generally the same method yields, for $\beta > \alpha$

$$\begin{aligned} I_\lambda &= \frac{1}{|\alpha\beta|} \int_{-\infty}^{+\infty} Ai\left[\frac{x+a}{\alpha}\right] Ai\left[\frac{x+b}{\beta}\right] \mathrm{e}^{\mathrm{i}\lambda x} \mathrm{d}x \qquad (3.126) \\ &= \frac{1}{\left(\beta^3 - \alpha^3\right)^{1/3}} \exp\left[-\mathrm{i}\frac{\lambda}{\beta^3 - \alpha^3}\left(\lambda^2\alpha^3\beta^3/3 + a\beta^3 - b\alpha^3\right)\right] \\ &\quad \times Ai\left[\frac{1}{\left(\beta^3 - \alpha^3\right)^{1/3}}\left(b - a - \frac{\lambda^2\alpha^3\beta^3}{\beta^3 - \alpha^3}\right)\right], \end{aligned}$$

and for $\beta = \alpha$

$$I_\lambda = \frac{1}{2\sqrt{\pi}\,|\lambda|^{1/2}\,|\alpha|^{3/2}} \times \exp\left\{-\mathrm{i}\left[\frac{\alpha^3\lambda^3}{12} - \frac{(a-b)^2}{4\alpha^3\lambda} + \frac{\lambda(a+b)}{2} + \frac{\pi}{4}\mathrm{sgn}(\alpha\lambda)\right]\right\}. \tag{3.127}$$

Finally, we give the following integral

$$\int_0^\infty Ai\left(\frac{x}{t}\right) Ai(t)\sqrt{t}\,\mathrm{d}t = \frac{-1}{2^{2/3}\sqrt{3}} Ai'(2^{2/3}\sqrt{x}), \; x > 0. \tag{3.128}$$

3.6 Other integrals

3.6.1 *Integrals involving the Volterra μ-function*

The Volterra μ-function is relatively intriguing in the bestiary of special functions, for it has received little attention, despite its remarkable properties. In this section, we would like to add some properties linked to Airy functions [Vallée (2002)].

In his work on integral equations with logarithmic kernel, Volterra introduced a function that can be generalised by Erdélyi *et al.* (1981)

$$\mu(u, \beta, \alpha) = \int_0^\infty \frac{t^\beta}{\Gamma(\beta+1)} \frac{u^{\alpha+t}}{\Gamma(t+\alpha+1)}\,\mathrm{d}t. \tag{3.129}$$

We are going to calculate the following integral by using the definition of the μ-function.

$$\int_0^\infty Ai(u)\,\mu(\gamma u, \beta, \alpha)\,\mathrm{d}u = \int_0^\infty \mathrm{d}t \frac{t^\beta}{\Gamma(\beta+1)} \int_0^\infty Ai(u) \frac{(\gamma u)^{\alpha+t}}{\Gamma(t+\alpha+1)}\,\mathrm{d}u. \tag{3.130}$$

Using the Mellin transform of the Airy function (see Eq. (3.86))

$$\int_0^\infty Ai(u)\,u^n\,\mathrm{d}u = \frac{\Gamma(n+1)}{3^{(n+3)/3}\Gamma(\frac{n+3}{3})}.$$

We then find, for the right member of Eq. (3.130),

$$\int_0^\infty \frac{\mathrm{d}t}{3} \frac{t^\beta}{\Gamma(\beta+1)} \frac{(\gamma^3/3)^{(\alpha+t)/3}}{\Gamma(\frac{\alpha+t}{3}+1)}.$$

Making the change $t = 3\,z$, we obtain

$$\int_0^\infty Ai(u)\,\mu(\gamma u, \beta, \alpha)\,\mathrm{d}u = 3^\beta \int_0^\infty \frac{z^\beta}{\Gamma(\beta+1)} \frac{(\gamma^3/3)^{\alpha/3+z}}{\Gamma(\frac{\alpha}{3}+z+1)}\,\mathrm{d}z,$$

from which we deduce the result

$$\int_0^\infty Ai(u)\,\mu(\gamma u,\beta,\alpha)\,\mathrm{d}u = 3^\beta \mu(\frac{\gamma^3}{3},\beta,\frac{\alpha}{3}). \tag{3.131}$$

This can also be written

$$\frac{1}{(3t)^{1/3}}\int_0^\infty Ai\left(\frac{x}{(3t)^{1/3}}\right)\,\mu(x,\beta,\alpha)\,\mathrm{d}x = 3^\beta \mu(t,\beta,\frac{\alpha}{3}). \tag{3.132}$$

Therefore, we have found that $\mu(x,\beta,0)$ is an eigenfunction of the above integral equation for the kernel

$$\frac{1}{(3t)^{1/3}} Ai\left(\frac{x}{(3t)^{1/3}}\right). \tag{3.133}$$

This kernel is easily recognised to satisfy (as a similarity solution) the evolution equation

$$\frac{\partial f(x,t)}{\partial t} + \frac{\partial^3 f(x,t)}{\partial x^3} = 0, \tag{3.134}$$

which is the linearised Korteweg–de Vries equation (l-KdV) [Ablowitz & Clarkson (1991)].

Volterra functions may be the eigenfunctions of many other integral equations involving the Airy function. By using the recurrence relations between Volterra functions

$$\begin{aligned} \tfrac{\mathrm{d}}{\mathrm{d}x}\mu(x,\beta,\alpha) &= \mu(x,\beta,\alpha-1) \\ x\mu(x,\beta,\alpha) &= \mu(x,\beta+1,\alpha+1) + (\alpha+1)\,\mu(x,\beta,\alpha+1). \end{aligned} \tag{3.135}$$

From the Airy equation, $Ai'' - u\,Ai = 0$, we can write

$$\int_0^\infty u\,Ai(u)\,\mu(\gamma u,\beta,\alpha)\,\mathrm{d}u = \int_0^\infty Ai''(u)\,\mu(\gamma u,\beta,\alpha)\,\mathrm{d}u. \tag{3.136}$$

We then perform an integration by parts

$$\begin{aligned} &\int_0^\infty Ai''(u)\,\mu(\gamma u,\beta,\alpha)\,\mathrm{d}u \\ &\qquad = Ai'(u)\,\mu(\gamma u,\beta,\alpha)|_0^\infty - \gamma\int_0^\infty Ai'(u)\,\mu'(\gamma u,\beta,\alpha)\,\mathrm{d}u \end{aligned} \tag{3.137}$$

The first term of the right member cancels, for $u = 0$ due to the μ-function and for $u \to \infty$ thanks to the Airy function. Now using the first of the relationships (3.135), we find

$$\int_0^\infty u\,Ai(u)\,\mu(\gamma u,\beta,\alpha)\,\mathrm{d}u = -\gamma\int_0^\infty Ai'(u)\,\mu(\gamma u,\beta,\alpha-1)\,\mathrm{d}u. \tag{3.138}$$

Making a second integration by parts

$$\int_0^\infty u\, Ai(u)\, \mu(\gamma u, \beta, \alpha)\, \mathrm{d}u = \gamma^2 \int_0^\infty Ai(u)\, \mu(\gamma u, \beta, \alpha - 2)\, \mathrm{d}u.$$

Thus using the result from Eq. (3.131), we obtain

$$\int_0^\infty u\, Ai(u)\, \mu(\gamma u, \beta, \alpha)\, \mathrm{d}u = 3^\beta \gamma^2\, \mu(\frac{\gamma^3}{3}, \beta, \frac{\alpha - 2}{3}). \tag{3.139}$$

There are other ways to prove this result. The first one is to take the derivative of Eq. (3.131)

$$\frac{\mathrm{d}}{\mathrm{d}\gamma} \int_0^\infty Ai(u)\, \mu(\gamma u, \beta, \alpha)\, \mathrm{d}u = \int_0^\infty u\, Ai(u)\, \mu(\gamma u, \beta, \alpha - 1)\, \mathrm{d}u,$$

which is equal to

$$\frac{\mathrm{d}}{\mathrm{d}\gamma} \left[3^\beta \mu(\frac{\gamma^3}{3}, \beta, \frac{\alpha}{3}) \right] = 3^\beta \gamma^2\, \mu(\frac{\gamma^3}{3}, \beta, \frac{\alpha}{3} - 1),$$

leading to the same result.

Another method is to use the recurrence relation between Volterra μ-functions (3.135). Eq. (3.139) may be written into the other form

$$\frac{1}{(3t)^{4/3}} \int_0^\infty x\, Ai\left[\frac{x}{(3t)^{1/3}}\right] \mu(x, \beta, \alpha)\, \mathrm{d}x = 3^\beta \gamma^2\, \mu(t, \beta, \frac{\alpha - 2}{3}). \tag{3.140}$$

Then $\mu(x, \beta, -1)$ is deduced to be the eigenfunction of an integral equation with the kernel

$$\frac{x}{(3t)^{4/3}} Ai\left(\frac{x}{(3t)^{1/3}}\right), \tag{3.141}$$

which is again a solution to Eq. (3.134).

In the same way, we can find other interesting results such as

$$\int_0^\infty Ai'(u)\, \mu(\gamma u, \beta, \alpha)\, \mathrm{d}u = -3^\beta \gamma\, \mu(\frac{\gamma^3}{3}, \beta, \frac{\alpha - 1}{3}), \tag{3.142}$$

and

$$\int_0^\infty Ai_1(u)\, \mu(\gamma u, \beta, \alpha)\, \mathrm{d}u = -3^\beta \frac{1}{\gamma}\, \mu(\frac{\gamma^3}{3}, \beta, \frac{\alpha + 1}{3}), \tag{3.143}$$

where $Ai_1(u) = \int_u^\infty Ai(t)\, \mathrm{d}t$.

More generally, we have a formula for the nth derivative of the Airy function

$$\int_0^\infty Ai^{(n)}(u)\, \mu(\gamma u, \beta, \alpha)\, \mathrm{d}u = 3^\beta (-\gamma)^n\, \mu(\frac{\gamma^3}{3}, \beta, \frac{\alpha - n}{3}). \tag{3.144}$$

Each of the equations (3.142) to (3.144), may be put into a form which satisfies an eigenfunction equation with a kernel that is a similarity solution to the l-KdV equation. Finally, we give a result for Ai^2 from its Mellin transform [Reid (1995)]

$$\int_0^\infty Ai^2(u)\, u^n \,\mathrm{d}u = \frac{2\Gamma(n+1)}{\sqrt{\pi}12^{(2n+7)/6}\Gamma(\frac{2n+7}{6})},$$

which reads

$$\int_0^\infty Ai^2(u)\, \mu(\gamma u, \beta, \alpha)\,\mathrm{d}u = \frac{3^\beta}{2\sqrt{\pi\gamma}}\,\mu(\frac{\gamma^3}{12}, \beta, \frac{2\alpha+1}{6}). \tag{3.145}$$

This equation may alternatively be written as

$$\frac{1}{(12t)^{1/6}}\int_0^\infty Ai^2\left[\frac{x}{(12t)^{1/3}}\right]\,\mu(x,\beta,\alpha)\,\mathrm{d}x = \frac{3^\beta}{2\sqrt{2}}\mu(t,\beta,\frac{2\alpha+1}{6}). \tag{3.146}$$

It is then found that $\mu(x, \beta, \frac{1}{4})$ is an eigenfunction of this equation, with the kernel

$$\frac{1}{(12t)^{1/6}}\,Ai^2\left[\frac{x}{(12t)^{1/3}}\right], \tag{3.147}$$

which again satisfies the l-KdV equation.

3.6.2 *Canonisation of cubic forms*

In this section, we shall derive a canonisation method of cubic forms, as described by Turnbull [Turnbull (1960)]. With the help of the following linear transformation

$$\begin{cases} X = \alpha x + \beta y \\ Y = \gamma x + \delta y \end{cases}$$

with

$$\Delta = \alpha\delta - \beta\gamma \neq 0.$$

We are looking for the transformation of the cubic form

$$F(x,y) = Ax^3 + 3Bx^2y + 3Cxy^2 + Dy^3$$

into its canonical form $F(X,Y) = X^3 + Y^3$. For this purpose, we consider the covariants and the invariant of the cubics

- The Hessian, i.e. the determinant of the second derivatives

$$\mathscr{H} = \left(AC - B^2\right)x^2 + (AD - BC)\,xy + \left(BD - C^2\right)y^2\ ;$$

- The cubic covariant is obtained by taking the Jacobian of the cubics and of the Hessian

$$\begin{aligned}\mathscr{C} = &\left(A^2D - 3ABC + 2B^3\right)x^3 + 3\left(ABD - 2AC^2 + B^2C\right)x^2y \\ &+3\left(2B^2D - ACD - BC^2\right)xy^2 + \left(3BCD - AD^2 - 2C^3\right)y^3.\end{aligned}$$

- The invariant of the cubics is built by taking the Hessian of the Hessian, which is the discriminant of the cubics F,

$$\mathscr{D} = (AD - BC)^2 - 4\left(AC - B^2\right)\left(BD - C^2\right).$$

There is a noteworthy relation (which is a so called *syzygy*) between the invariant and the covariants

$$F^2\mathscr{D} = \mathscr{C}^2 + 4\mathscr{H}^3.$$

For similar quantities of the canonical cubics $F(X,Y) = X^3 + Y^3$, we have

$$\begin{cases} \mathscr{H} = \Delta^2 XY, \\ \mathscr{C} = \Delta^3\left(X^3 - Y^3\right), \\ \mathscr{D} = \Delta^6. \end{cases}$$

By comparing the cubics F and the cubic covariant $\mathscr{C}$, identifying the terms x^3 and y^3 with the terms X^3 and Y^3 respectively, we finally obtain

$$\begin{cases} \alpha^3 = \frac{1}{2}\left[A + \frac{1}{\sqrt{\mathscr{D}}}(A^2D - 3ABC + 2B^3)\right], \\ \gamma^3 = \frac{1}{2}\left[A - \frac{1}{\sqrt{\mathscr{D}}}(A^2D - 3ABC + 2B^3)\right], \\ \beta^3 = \frac{1}{2}\left[D - \frac{1}{\sqrt{\mathscr{D}}}(AD^2 - 3BCD + 2C^3)\right], \\ \delta^3 = \frac{1}{2}\left[D + \frac{1}{\sqrt{\mathscr{D}}}(AD^2 - 3BCD + 2C^3)\right], \end{cases}$$

with the condition

$$\mathscr{D} = (AD - BC)^2 - 4\left(AC - B^2\right)\left(BD - C^2\right) = \Delta^6 \neq 0.$$

It should be noted, in the case where $\mathscr{D} = 0$, that one cannot use the general canonical form $F(X,Y) = X^3 + Y^3$. In this case, the canonical form of $F(x,y)$ is either X^3 or X^2Y.

3.6.3 *Integrals with three Airy functions*

Consider the following integral (where a, c and e are all non-zero) [Vallée (1982)]

$$L = \int_{-\infty}^{+\infty} Ai(ax+b)Ai(cx+d)Ai(ex+f)\mathrm{d}x. \tag{3.148}$$

Once again using the integral representation (2.21) of the Airy function

$$L = \frac{1}{8\pi^3} \iiiint\limits_{\mathbb{R}^4} \mathrm{e}^{\mathrm{i}(u^3+v^3+w^3)/3} \mathrm{e}^{\mathrm{i}(ax+b)u} \times \mathrm{e}^{\mathrm{i}(cx+d)v} \mathrm{e}^{\mathrm{i}(ex+f)w} \mathrm{d}u\mathrm{d}v\mathrm{d}w\mathrm{d}x.$$

An integration on the variable x results in a Dirac delta function allowing one to express w as a function of u and v

$$L = \frac{1}{4\pi^2} \iiint\limits_{\mathbb{R}^3} \mathrm{e}^{\mathrm{i}(u^3+v^3+w^3)/3} \mathrm{e}^{\mathrm{i}(bu+dv+fw)} \delta(au+cv+ew) \mathrm{d}u\mathrm{d}v\mathrm{d}w.$$

We then have an integral of the kind

$$L = \frac{1}{4\pi^2 \, |e|} \iint\limits_{\mathbb{R}^2} \mathrm{e}^{\mathrm{i}\left(Ax^3/3+Bx^2y+Cxy^2+Dy^3/3+Vx+Wy\right)} \mathrm{d}x\mathrm{d}y,$$

with

$$\begin{cases} A = 1 - a^3/\mathrm{e}^3, \\ B = -a^2c/\mathrm{e}^3, \\ C = -ac^2/\mathrm{e}^3, \\ D = 1 - c^3/\mathrm{e}^3, \\ V = b - af/e, \\ W = d - cf/e. \end{cases}$$

We consider here that the quantity

$$K = e^6 + a^6 + c^6 - 2a^3c^3 - 2a^3e^3 - 2c^3e^3 > 0.$$

The calculation of this integral is then carried out in §3.6.5, formula (3.154), and the result reads

$$L = \frac{1}{K^{1/6}} Ai\left(\frac{(be-af)\delta - (de-cf)\gamma}{K^{1/6}}\right) Ai\left(\frac{(de-cf)\alpha - (be-af)\beta}{K^{1/6}}\right),$$

with

$$\begin{cases} \alpha^3 = \frac{1}{2}\left[A + \frac{1}{\sqrt{K}}(A^2D - 3ABC + 2B^3)\right], \\ \beta^3 = \frac{1}{2}\left[D - \frac{1}{\sqrt{K}}(AD^2 - 3BCD + 2C^3)\right], \\ \gamma^3 = \frac{1}{2}\left[A - \frac{1}{\sqrt{K}}(A^2D - 3ABC + 2B^3)\right], \\ \delta^3 = \frac{1}{2}\left[D + \frac{1}{\sqrt{K}}(AD^2 - 3BCD + 2C^3)\right]. \end{cases}$$

If $K < 0$ the integral will involve the irregular companion Bi of the Airy function Ai. As an example, let us consider the integral

$$M = \int_{-\infty}^{+\infty} Ai(x)Ai(x+a)Ai(x+b)\mathrm{d}x$$

The calculation of this integral proceeds as follows. We first use the integral representation of the product of two Airy functions (cf. formula (2.153))

$$Ai(x)Ai(x+a) = \frac{1}{2^{1/3}\pi}\int_{-\infty}^{+\infty} Ai\left(\frac{x+t^2+a/2}{2^{-2/3}}\right) \mathrm{e}^{iat}\,\mathrm{d}t.$$

So we have

$$M = \frac{1}{2^{1/3}\pi}\int_{-\infty}^{+\infty} \mathrm{e}^{iat}\,\mathrm{d}t \int_{-\infty}^{+\infty} Ai\left(\frac{x+t^2+a/2}{2^{-2/3}}\right) Ai(x+b)\mathrm{d}x$$

leading, thanks to Eq. (3.111), to

$$M = \frac{1}{2\pi}\left(\frac{4}{3}\right)^{1/3}\int_{-\infty}^{+\infty} Ai\left[\left(\frac{4}{3}\right)^{1/3}(b-a/2-t^2)\right]\mathrm{e}^{iat}\,\mathrm{d}t.$$

We then use the non-trivial result, formula (3.125), to obtain

$$M = 3^{-1/6}\Re\left\{Ai\left(\frac{d+\mathrm{i}c}{2^{2/3}}\right) Bi\left(\frac{d-\mathrm{i}c}{2^{2/3}}\right)\right\},$$

with the notation $c=(\frac{4}{3})^{1/3}(b-a/2)$ and $\mathrm{d}=(\frac{4}{3})^{1/6}a$, and $\Re$, the real part. A little algebra yields

$$M = 3^{-1/6}\Re\left\{Ai\left[3^{-1/3}(b+\mathrm{j}a)\right]\, Bi\left[3^{-1/3}(b+\mathrm{j}^2a)\right]\right\}, \tag{3.149}$$

where j and j^2 are the cubic roots of unity.

We give another particular case of an integral with the product of three Airy functions [Vallée *et al.* (1997)]

$$\begin{aligned}\int_{-\infty}^{+\infty} & Ai(x)Ai(y-x)Ai(z-x)\mathrm{d}x \\ &= 5^{-1/6}Ai\left[5^{-1/3}\left(y/\varepsilon-\varepsilon z\right)\right] Ai\left[5^{-1/3}\left(z/\varepsilon-\varepsilon y\right)\right]\end{aligned} \tag{3.150}$$

where ε is the golden mean $\varepsilon=\frac{\sqrt{5}-1}{2}$.

3.6.4 *Integrals with four Airy functions*

Let us consider the integral [Aspnes (1966)]

$$I = \int_{-\infty}^{+\infty} Ai^2\left(\frac{a+x}{\alpha}\right) Ai^2\left(\frac{b-x}{\beta}\right)\mathrm{d}x \tag{3.151}$$

with $\alpha > 0$ and $\beta > 0$. With the help of the formula (2.157), this integral reads

$$I = \frac{1}{2^{4/3}\pi^2} \int_{-\infty}^{+\infty} \mathrm{d}x \int_0^{\infty} \frac{\mathrm{d}u}{\sqrt{u}} \int_0^{\infty} \frac{\mathrm{d}v}{\sqrt{v}} Ai\left[\frac{2^{2/3}}{\alpha}(a+x)+u\right] \times Ai\left[\frac{2^{2/3}}{\beta}(b-x)+v\right].$$

By now using the formula (3.111), we have

$$I = \frac{\alpha\beta}{4\pi^2\left(\alpha^3+\beta^3\right)^{1/3}} \int_0^{\infty} \frac{\mathrm{d}u}{\sqrt{u}} \int_0^{\infty} \frac{\mathrm{d}v}{\sqrt{v}}\, Ai\left[\frac{2^{2/3}(a+b)+\alpha u+\beta v}{\left(\alpha^3+\beta^3\right)^{1/3}}\right].$$

With the help of formula (2.157), this expression becomes

$$I = \frac{\sqrt{\alpha\beta}}{2\pi} \int_0^{\infty} \frac{\mathrm{d}w}{\sqrt{w}}\, Ai^2\left[\frac{(a+b)}{\left(\alpha^3+\beta^3\right)^{1/3}}+w\right].$$

We find here the primitive of the Airy function Ai (cf. formula (3.108))

$$Ai_1(x) = 2\int_0^{\infty} Ai^2\left(t+2^{-2/3}x\right)\frac{\mathrm{d}t}{\sqrt{t}}.$$

We finally obtain

$$\int_{-\infty}^{+\infty} Ai^2\left(\frac{a+x}{\alpha}\right) Ai^2\left(\frac{b-x}{\beta}\right)\mathrm{d}x = \frac{\sqrt{\alpha\beta}}{4\pi} Ai_1\left[2^{2/3}\frac{(a+b)}{\left(\alpha^3+\beta^3\right)^{1/3}}\right]. \tag{3.152}$$

3.6.5 *Double integrals*

Let us consider the following double integral [Vallée (1982)]

$$G = \iint_{\mathbb{R}^2} Ai(ax+by+p)Ai(cx+dy+q) \times Ai(ex+fy+r)Ai(gx+hy+w)\mathrm{d}x\mathrm{d}y, \tag{3.153}$$

where a, c, e, g are non-zero constants. Making use of the integral representation (2.21) of $Ai(x)$ and rearranging the terms, produces

$$G = \frac{1}{16\pi^4} \iiiint_{\mathbb{R}^4} \mathrm{d}t\mathrm{d}t'\mathrm{d}s\mathrm{d}s'\ \mathrm{e}^{\mathrm{i}\left[\left(t^3+t'^3+s^3+s'^3\right)/3+pt+qt'+rs+ws'\right]} \times \iint_{\mathbb{R}^2} \mathrm{e}^{\mathrm{i}\left[x\left(at+ct'+es+gs'\right)+y\left(bt+dt'+fs+hs'\right)\right]}\mathrm{d}x\mathrm{d}y.$$

From the integral representation of the Dirac delta function $\delta(z) = \frac{1}{2\pi}\int_{-\infty}^{+\infty} \mathrm{e}^{\mathrm{i}zw}\mathrm{d}w$, we have

$$G = \frac{1}{16\pi^4} \iiiint_{\mathbb{R}^4} \mathrm{d}t\mathrm{d}t'\mathrm{d}s\mathrm{d}s'\ \mathrm{e}^{\mathrm{i}\left[\left(t^3+t'^3+s^3+s'^3\right)/3+pt+qt'+rs+ws'\right]}$$
$$\times\ \delta\left(at+ct'+es+gs'\right)\delta\left(bt+dt'+fs+hs'\right).$$

An integration over s and s', then gives us

$$G = \frac{K^3}{4\pi^2} \iint_{\mathbb{R}^2} \mathrm{d}x\mathrm{d}y\ \mathrm{e}^{\mathrm{i}\left(Ax^3/3+Bx^2y+Cxy^2+Dy^3/3+Vx+Wy\right)}, \tag{3.154}$$

with

$$\begin{cases} K = eh - fg \neq 0, \\ A = (eh-fg)^3 + (bg-ah)^3 + (af-be)^3, \\ D = (eh-fg)^3 + (dg-ch)^3 + (cf-de)^3, \\ B = (bg-ah)^2(dg-ch) + (af-be)^2(cf-de), \\ V = v(bg-ah) + w(af-be) + p(eh-fg), \\ W = v(dg-ch) + w(cf-de) + q(eh-fg). \end{cases}$$

The next step is to canonise the cubic form (cf. §3.6.2), leading to

$$G = \frac{K^3}{4\pi^2\Delta} \iint_{\mathbb{R}^2} \mathrm{d}X\mathrm{d}Y\ \exp\left[\mathrm{i}\left(\frac{X^3}{3}+\frac{Y^3}{3}\right.\right.$$
$$\left.\left.+\frac{V\delta-W\gamma}{\Delta}X+\frac{W\alpha-V\beta}{\Delta}\right)Y\right], \tag{3.155}$$

with

$$\Delta^6 = \mathscr{D} = (AD-BC)^2 - 4\left(AC-B^2\right)\left(BD-C^2\right) > 0.$$

Finally, the integral (3.153) may be written as

$$G = \frac{K^3}{\Delta} Ai\left(\frac{V\delta-W\gamma}{\Delta}\right) Ai\left(\frac{W\alpha-V\beta}{\Delta}\right). \tag{3.156}$$

Exercises

(1) Generalise the table of Albright (see §3.2.1) to calculate the following primitives (n=1, 2)

$$\int x^n y_1^2\,\mathrm{d}x, \quad \int x^n y_1 y\,\mathrm{d}x, \quad \int x^n y_1 y'\,\mathrm{d}x.$$

(2) Calculate the following integrals in terms of Ai and Gi functions

$$\int_0^{\infty} \cos\left(a\,t^3 + b\,t + c\right)\,\mathrm{d}t, \quad \int_0^{\infty} \sin\left(a\,t^3 + b\,t + c\right)\,\mathrm{d}t.$$

(3) Calculate

$$\int_0^{\infty} \cos\left(\frac{t^3}{12} + \alpha t - \frac{\beta}{t} + \frac{\pi}{4}\right)\frac{\mathrm{d}t}{\sqrt{t}}.$$

(4) Use the relations between Bessel and Airy functions (cf. §2.2.4) to express the following integral in terms of Airy functions

$$\int_0^{\infty} x\mathrm{e}^{-\frac{x^2}{2a}}\left[I_\nu(x) + I_{-\nu}(x)\right]K_\nu(x)\,\mathrm{d}x = a\mathrm{e}^{a}\,K_\nu(a),$$

when $\nu = \frac{1}{3}$ and $\nu = \frac{2}{3}$, a > 0.

(5) Check the formula (3.150), with the method described in §3.6.3.

Chapter 4

Transformations of Airy Functions

4.1 Causal properties of Airy functions

4.1.1 *Causal relations*

Using the integral representation of $Ai(x)$, $Gi(x)$, $Ai^2(x)$ and $Ai(x)Bi(x)$ given by the formulae (2.21), (2.127), (2.150) and (2.151) respectively, we can write [Scorer (1950); Aspnes (1967)]

$$Ai(x) + \mathrm{i}Gi(x) = \frac{1}{\pi} \int_0^{\infty} \mathrm{e}^{\mathrm{i}\left(t^3/3+tx\right)} \mathrm{d}t \tag{4.1}$$

$$Ai^2(x) + \mathrm{i}Ai(x)Bi(x) = \frac{1}{2\pi^{3/2}} \int_0^{\infty} \mathrm{e}^{\mathrm{i}\left(t^3/12+tx+\pi/4\right)} \frac{\mathrm{d}t}{\sqrt{t}}. \tag{4.2}$$

Being analytic in the complex plane, the real and imaginary parts of these functions may be written as a Hilbert transform. We have then

$$Ai(x) = \frac{1}{\pi} \wp \int_{-\infty}^{+\infty} \frac{Gi(x')}{x'-x} \mathrm{d}x', \tag{4.3}$$

and conversely

$$Gi(x) = -\frac{1}{\pi} \wp \int_{-\infty}^{+\infty} \frac{Ai(x')}{x'-x} \mathrm{d}x', \tag{4.4}$$

where $\wp$ is the Cauchy principal value. From the relation (4.2), we can also deduce

$$Ai^2(x) = \frac{1}{\pi} \wp \int_{-\infty}^{+\infty} \frac{Ai(x')Bi(x')}{x'-x} \mathrm{d}x', \tag{4.5}$$

and conversely

$$Ai(x)Bi(x) = -\frac{1}{\pi}\wp\int_{-\infty}^{+\infty}\frac{Ai^2(x')}{x'-x}\mathrm{d}x'. \quad (4.6)$$

Using the relation (4.1), we can write

$$\frac{1}{|\alpha|}\left[Ai\left(\frac{x}{\alpha}\right)+\mathrm{i}Gi\left(\frac{x}{\alpha}\right)\right] = \frac{1}{\pi}\int_0^{\infty}\mathrm{e}^{\mathrm{i}(\alpha^3u^3/3+ux)}\mathrm{d}u. \quad (4.7)$$

So that when $\alpha \to 0$, this last relation becomes

$$\lim_{\alpha\to 0}\frac{1}{|\alpha|}\left[Ai\left(\frac{x}{\alpha}\right)+\mathrm{i}Gi\left(\frac{x}{\alpha}\right)\right] = \frac{1}{\pi}\int_0^{\infty}\mathrm{e}^{\mathrm{i}ux}\mathrm{d}u = \delta(x)+\mathrm{i}\frac{1}{\pi}\wp\frac{1}{x},$$

that is to say

$$\lim_{\alpha\to 0}\frac{1}{|\alpha|}Ai\left(\frac{x}{\alpha}\right) = \delta(x), \quad (4.8)$$

and

$$\lim_{\alpha\to 0}\frac{1}{|\alpha|}Gi\left(\frac{x}{\alpha}\right) = \frac{1}{\pi}\wp\frac{1}{x}. \quad (4.9)$$

It can be seen that these last two expressions are the beginning of the expansions given by Lee (1980):

$$\frac{1}{|\alpha|}Ai\left(\frac{x}{\alpha}\right) = \delta(x)+\frac{\alpha^3}{3}\delta^{(3)}(x)-\frac{\alpha^6}{18}\delta^{(6)}(x)+\dots \quad (4.10)$$

and

$$\frac{1}{|\alpha|}Gi\left(\frac{x}{\alpha}\right) \quad (4.11)$$
$$= \frac{1}{\pi}\left[\wp\left(\frac{1}{x}\right)+\frac{\alpha^3}{3}\frac{\mathrm{d}^3}{\mathrm{d}x^3}\wp\left(\frac{1}{x}\right)-\frac{\alpha^6}{18}\frac{\mathrm{d}^6}{\mathrm{d}x^6}\wp\left(\frac{1}{x}\right)+\dots\right].$$

4.1.2 *Green's function of the Airy equation*

In this section, we are going to build the Green's function satisfying the differential equation [Moyer (1973); Burnett & Belsley (1983)]

$$\left(\frac{\partial^2}{\partial x^2}-x\right)G(x,x') = \delta(x-x'). \quad (4.12)$$

The integral expression of this function, given by Lukes & Somaratna (1969), is

$$G(x,x') = -\mathrm{i}\int_0^\infty U(x,x',t)\mathrm{d}t, \tag{4.13}$$

with

$$U(x,x',t) = \left(\frac{1}{4\pi t}\right)^{1/2} \exp\left\{\mathrm{i}\left[-\frac{\pi}{4} + \left(\frac{x-x'}{4}\right)^2 \frac{1}{t} - \left(\frac{x+x'}{2}\right)t - \frac{t^3}{12}\right]\right\}.$$

We can write, thanks to the integral representation (2.21) of the function $Ai(x)$,

$$Ai\left(z'\mathrm{e}^{\mathrm{i}2\pi/3}\right) Ai\left(z\right) = \frac{\mathrm{e}^{\mathrm{i}\pi/12}}{4\pi^{3/2}} \int_0^\infty \mathrm{e}^{\mathrm{i}\left[-\left(\frac{z+z'}{2}\right)v + \left(\frac{z-z'}{2}\right)^2 \frac{1}{v} - \frac{v^3}{12}\right]} \frac{\mathrm{d}v}{\sqrt{v}}, \tag{4.14}$$

for $0 \leq \arg(z-z') \leq \pi/2$, and

$$Ai\left(z\mathrm{e}^{\mathrm{i}2\pi/3}\right) Ai\left(z'\right) = \frac{\mathrm{e}^{\mathrm{i}\pi/12}}{4\pi^{3/2}} \int_0^\infty \mathrm{e}^{\mathrm{i}\left[-\left(\frac{z+z'}{2}\right)v + \left(\frac{z-z'}{2}\right)^2 \frac{1}{v} - \frac{v^3}{12}\right]} \frac{\mathrm{d}v}{\sqrt{v}}, \tag{4.15}$$

for $0 \leq \arg(z'-z) \leq \pi/2$. Comparing Eqs. (4.13) to (4.15), we obtain

$$G(x,x') = 2\pi\mathrm{i}\mathrm{e}^{\mathrm{i}2\pi/3} \times \begin{cases} Ai\left(x'\mathrm{e}^{\mathrm{i}2\pi/3}\right) Ai\left(x\right) \text{ if } x \geq x' \\ \\ Ai\left(x'\right) Ai\left(x\mathrm{e}^{\mathrm{i}2\pi/3}\right) \text{ if } x \leq x'. \end{cases}$$

Thanks to the relation (see formula (2.15))

$$\mathrm{e}^{\mathrm{i}2\pi/3} Ai\left(z\mathrm{e}^{\mathrm{i}2\pi/3}\right) = -\frac{1}{2}\left[Ai(z) - \mathrm{i}Bi(z)\right],$$

$G(x,x')$ can be written as a function of Ai and Bi

$$G(x,x') = -\pi \times \begin{cases} Ai(x)Bi(x') + \mathrm{i}Ai(x)Ai(x') & \text{if } x \geq x' \\ \\ Ai(x')Bi(x) + \mathrm{i}Ai(x')Ai(x) & \text{if } x \leq x'. \end{cases}$$

4.1.3 *Fractional derivatives of Airy functions*

In a series of papers, Varlamov [Varlamov (2007), (2008-a,b,c); Temme & Varlamov (2009)] was particularly interested in the fractional derivatives of Airy functions and their applications.

In order to give a rapid overview of this subject, we first introduce a definition of fractional derivatives. For real $\alpha > -1$, the fractional derivative of order α of a function f is defined by the relation

$$\mathrm{D}^\alpha f(x) = \mathrm{D}^\alpha\{f(x)\} = \frac{1}{2\pi}\int_{-\infty}^{\infty} |\xi|^\alpha \hat{f}(\xi)\mathrm{e}^{\mathrm{i}\xi x}\,\mathrm{d}\xi, \tag{4.16}$$

subject to the convergence of the integral and where $\hat{f}$ is the Fourier transform of f. We give now several results established by Varlamov. He first introduces some functions which are linear combinations of products of Airy functions

$$w_+(x) = Ai(x)Bi(x) + Ai^2(x), \tag{4.17}$$

and

$$w_-(x) = Ai(x)Bi(x) - Ai^2(x). \tag{4.18}$$

He then proves the following relations

$$\mathrm{D}^{-1/2}Gi(x) = \kappa w_-\left(\frac{x}{2^{2/3}}\right), \tag{4.19}$$

$$\mathrm{D}^{-1/2}Ai(x) = \kappa w_+\left(\frac{x}{2^{2/3}}\right), \tag{4.20}$$

where $\kappa = 2^{1/6}\sqrt{\pi}$. Moreover

$$\mathrm{D}^{1/2}Ai(x) = \frac{\kappa}{2^{2/3}}w'_-\left(\frac{x}{2^{2/3}}\right), \tag{4.21}$$

$$\mathrm{D}^{1/2}Gi(x) = \frac{-\kappa}{2^{2/3}}w'_+\left(\frac{x}{2^{2/3}}\right). \tag{4.22}$$

More generally, we have for $\alpha > -1/2$

$$\mathrm{D}^\alpha[Ai^2(x)] = k_\alpha[\mathrm{D}^{\alpha-1/2}Ai(2^{2/3}x) - \mathrm{D}^{\alpha-1/2}Gi(2^{2/3}x)], \tag{4.23}$$

and

$$\mathrm{D}^\alpha[Ai(x)Bi(x)] = k_\alpha[\mathrm{D}^{\alpha-1/2}Ai(2^{2/3}x) + \mathrm{D}^{\alpha-1/2}Gi(2^{2/3}x)], \tag{4.24}$$

where $k_\alpha = \frac{2^{2(\alpha-1)/3}}{\sqrt{2\pi}}$.

From these results some very nice relationships may be found. Let

$$d_1 = \frac{1}{k_\alpha}\mathrm{D}^\alpha\big(Ai(x)Bi(x)\big), \quad d_2 = \frac{1}{k_\alpha}\mathrm{D}^\alpha\big(Ai^2(x)\big),$$

$$a = \left(\mathrm{D}^{\alpha-1/2}Ai\right)\left(2^{2/3}x\right), \quad b = \left(\mathrm{D}^{\alpha-1/2}Gi\right)\left(2^{2/3}x\right). \tag{4.25}$$

Then we have the following relationships [Varlamov (2008-b)]

$$d_1^2 + d_2^2 = 2(a^2 + b^2), \quad d_1^2 - d_2^2 = 4ab \quad \text{and} \quad d_1 d_2 = a^2 - b^2. \tag{4.26}$$

The results given in §(3.3) are generalised by using Eq. (3.70). Varlamov (2008-b) obtains the non-trivial result

$$\int \frac{\mathrm{D}^{\alpha-1/2}Ai(x).\mathrm{D}^{\alpha+1/2}Ai(x) + \mathrm{D}^{\alpha-1/2}Gi(x).\mathrm{D}^{\alpha+1/2}Gi(x)}{[\mathrm{D}^{\alpha-1/2}Ai(x) + \mathrm{D}^{\alpha-1/2}Gi(x)]^2}$$
$$\times f\left(\frac{\mathrm{D}^{\alpha-1/2}Ai(x) + \mathrm{D}^{\alpha-1/2}Gi(x)}{\mathrm{D}^{\alpha-1/2}Ai(x) - \mathrm{D}^{\alpha-1/2}Gi(x)}\right) \mathrm{d}x = 2^{2/3} F\left(\frac{\mathrm{D}^{\alpha}[Ai(y)Bi(y)]}{\mathrm{D}^{\alpha}[Ai^2(y)]}\right),$$

where $y = 2^{-2/3}x$ and $F' = f$, from which many particular cases may be found. Several other results concerning the fractional derivatives of Airy functions can be found in the work by Varlamov.

4.2 The Airy transform

4.2.1 *Definitions and elementary properties*

We define the family of functions

$$\omega_\alpha(x) = \frac{1}{|\alpha|} Ai\left(\frac{x}{\alpha}\right), \quad \alpha \in \mathbb{R}. \tag{4.27}$$

One of the most important properties of this family is (see formula (4.8))

$$\lim_{\alpha \to 0} \{\omega_\alpha(x)\} = \delta(x), \tag{4.28}$$

where $\delta(x)$ is the Dirac delta function and $\alpha \in \mathbb{R}$. Moreover, from formula (3.111), we can set the relation giving the convolution product of two functions $\omega_\alpha(x)$ and $\omega_\beta(x)$

$$\omega_\alpha * \omega_\beta(x) = \omega_\gamma(x), \tag{4.29}$$

with $\gamma^3 = \alpha^3 + \beta^3$. In other words, the family of functions defined by Eq. (4.27) is stable under the convolution product and then leads to a semigroup of convolution. Note also that the functions ω_α are normalised (formula (3.76))

$$\int_{-\infty}^{+\infty} Ai(x)\mathrm{d}x = \int_{-\infty}^{+\infty} \omega_\alpha(x)\mathrm{d}x = 1. \tag{4.30}$$

Therefore if f is a function of x, and $\hat{f}$ its Fourier transform, we write φ_α, the Airy transform of f, as the convolution product

$$\varphi_\alpha(x) = f * \omega_\alpha(x) = \mathcal{A}_\alpha \left[f(x) \right], \tag{4.31}$$

which is also written

$$\varphi_\alpha(x) = \frac{1}{2\pi} \int_{-\infty}^{+\infty} \mathrm{e}^{\mathrm{i}(\alpha^3 \xi^3/3 + \xi x)} \hat{f}(\xi) \mathrm{d}\xi. \tag{4.32}$$

Consequently, $f \mapsto \varphi_\alpha$ is a particular class of functional transform that can be reversed by the formula

$$f(x) = \varphi_\alpha * \omega_{-\alpha}(x) = \overline{\mathcal{A}}_\alpha \left[\varphi_\alpha(x) \right]. \tag{4.33}$$

In order to sum up the results, Eqs. (4.31) and (4.33) define the *Airy transform*, where the inverse satisfies $\mathcal{A}_\alpha \overline{\mathcal{A}}_\alpha = \mathcal{A}_\alpha \mathcal{A}_{-\alpha} = \mathcal{I}$.

Physical applications of this transform can be found in the work by Hunt [Hunt (1981)] concerning molecular physics, and a work [Bertoncini *et al.* (1989, 1990)] using the Airy transform as a tool in the calculation of the Green's function of non-equilibrium high field quantum transport.

In fact, the Airy transform was mathematically introduced by Widder [Widder (1979)] from another approach i.e. by considering the self-adjoint Schrödinger operator: $\left\{ x - \frac{\mathrm{d}^2}{\mathrm{d}x^2} \right\}$. The solutions of the eigenvalue equation are $Ai(\xi - x)$. These functions form a continuous set of eigenfunctions, with the eigenvalues ξ. Indeed, from Eq. (3.111), we have the orthogonality condition

$$\int_{-\infty}^{+\infty} Ai(\xi - x) Ai(\xi' - x) dx = \delta(\xi - \xi').$$

The definition of the Airy transform then follows

$$\varphi(\xi) = \int_{-\infty}^{+\infty} f(x) Ai(\xi - x) \mathrm{d}x, \tag{4.34}$$

and the inverse transform

$$f(x) = \int_{-\infty}^{+\infty} \varphi(\xi) Ai(\xi - x) \mathrm{d}\xi. \tag{4.35}$$

Therefore, in the transformation defined by Eqs. (4.31) and (4.33), α appears as a scaling parameter, in relation to the basic transform defined by the above Eqs. (4.34) and (4.35), but in this case with $\alpha > 0$.

From the definition of the Airy transform, some elementary properties can be derived:

Proposition 4.1. ***Translation***
If $\varphi_\alpha(x)$ is the transform of $f(x)$, then $\varphi_\alpha(x+s)$ is the transform of $f(x+s)$.

Proposition 4.2. ***Scaling***
If $\varphi_\alpha(x)$ is the transform of $f(x)$, then $\varphi_{\alpha k}(kx)$ is the transform of $f(kx)$.

Proposition 4.3. ***Derivative***
If $\varphi_\alpha(x)$ is the transform of $f(x)$, then $\varphi'_\alpha(x)$ is the transform of $f'(x)$.

Proposition 4.4. ***Iteration***
If $\varphi_\alpha(x)$ is the transform of $f(x)$ by the function $\omega_\alpha(x)$, then the transform of $\varphi_\alpha(x)$ by the function $\omega_\beta(x)$ is $\varphi_\gamma(x)$, where $\gamma^3 = \alpha^3 + \beta^3$. In other words

$$\varphi_\gamma = (f * \omega_\alpha) * \omega_\beta = f * (\omega_\alpha * \omega_\beta) = f * \omega_\gamma. \tag{4.36}$$

Proposition 4.5. ***Convolution***
If $\varphi_\alpha(x)$ is the transform of $f(x)$ by the function $\omega_\alpha(x)$ and $\psi_\beta(x)$ is the transform of $g(x)$ by the function $\omega_\beta(x)$, then the convolution product $\varphi_\alpha * \psi_\beta$ is the transform of $f * g$ by the function $\omega_\gamma(x)$, where $\gamma^3 = \alpha^3 + \beta^3$

$$\varphi_\alpha * \psi_\beta = (f * \omega_\alpha) * (g * \omega_\beta) = (f * g) * (\omega_\alpha * \omega_\beta) = (f * g) * \omega_\gamma. \tag{4.37}$$

As the Airy functions $\{Ai(\xi - x),\ \xi \in \mathbb{R}\}$ form a continuous basis of orthogonal functions, we have:

Proposition 4.6. ***The Plancherel–Parseval rule***
If $\varphi_\alpha(x)$ and $\psi_\alpha(x)$ are the transforms of the real functions $f(x)$ and $g(x)$ by $\omega_\alpha(x)$, then for all real α

$$\int_{-\infty}^{+\infty} f(x)g(x)\mathrm{d}x = \int_{-\infty}^{+\infty} \varphi_\alpha(x)\psi_\alpha(x)\mathrm{d}x. \tag{4.38}$$

The proof of all these properties of the Airy transform can be shown rigorously, with the methods described in the paper by Widder. The same holds for the examples presented in the next section.

4.2.2 *Some examples*

Let us start with the transform of a constant. From Eq. (4.30), we immediately deduce that the transform of $f(x) = 1$ is $\varphi_\alpha(x) = 1$.

From this first case, we can examine the periodic functions by putting

$$f(x) = \mathrm{e}^{\mathrm{i}\xi x}.$$

We then obtain, with the integral representation (2.21) of Ai and the Dirac delta representation,

$$\varphi_\alpha(x) = \frac{1}{|\alpha|}\int_{-\infty}^{+\infty} \mathrm{e}^{\mathrm{i}\xi y} Ai\left(\frac{x-y}{\alpha}\right) \mathrm{d}y = \mathrm{e}^{\mathrm{i}\left(\alpha^3\xi^3/3+\xi x\right)}.$$

Now from this equation, we can obtain the transform of periodic functions

$$\sin(\xi x) \xrightarrow{\mathcal{A}_\alpha} \sin\left(\xi x + \frac{\alpha^3\xi^3}{3}\right) \tag{4.39}$$

$$\cos(\xi x) \xrightarrow{\mathcal{A}_\alpha} \cos\left(\xi x + \frac{\alpha^3\xi^3}{3}\right). \tag{4.40}$$

This result may be employed to transform a periodic function of which we have the Fourier expansion

$$f(x) = \sum c_n \mathrm{e}^{\mathrm{i}\pi n x}.$$

The Airy transform of this function is then

$$\varphi_\alpha(x) = \sum b_n(\alpha)\mathrm{e}^{\mathrm{i}\pi n x},$$

with

$$b_n = c_n \mathrm{e}^{\mathrm{i}(\pi n \alpha)^3}.$$

We now consider the case of a normalised Gaussian function

$$f(x) = \frac{1}{\sqrt{\pi}}\,\mathrm{e}^{-x^2}. \tag{4.41}$$

A little algebra gives the result of the transform

$$\varphi_\alpha(x) = \frac{1}{|\alpha|}\,\mathrm{e}^{\frac{1}{4\alpha^3}\left(x+\frac{1}{24\alpha^3}\right)} Ai\left(\frac{x}{\alpha} + \frac{1}{16\alpha^4}\right). \tag{4.42}$$

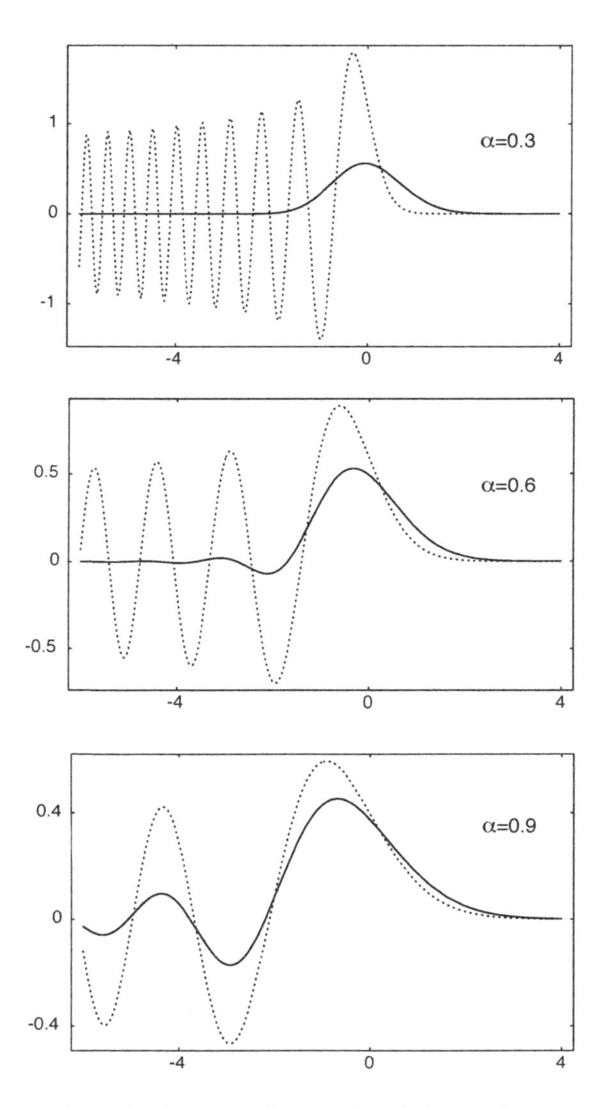

Fig. 4.1 Airy transform of a Gaussian: the transform (solid lines) given from the formula (4.42) is compared to the normalized Airy function $\omega_\alpha(x)$ (formula (4.27)), for $\alpha = 0.3$, $\alpha = 0.6$, $\alpha = 0.9$.

On Fig. (4.1), the Airy function $\omega_\alpha(x)$ (dotted lines) and the transform of the Gaussian function $\varphi_\alpha(x)$ (solid lines) are plotted for different values of the parameter α. For the smaller values of α, the transform resembles a

Gaussian but its behaviour rapidly tends to that of an Airy function, when α increases, losing its Gaussian character, even asymptotically.

The last example concerns the step function $\theta(x)$. Its Airy transform is

$$\varphi_\alpha(x) = \frac{1}{|\alpha|} \int_{-\infty}^{+\infty} \theta(y) Ai\left(\frac{x-y}{\alpha}\right) \mathrm{d}y = \int_{-\infty}^{x/\alpha} Ai(u)\mathrm{d}u. \tag{4.43}$$

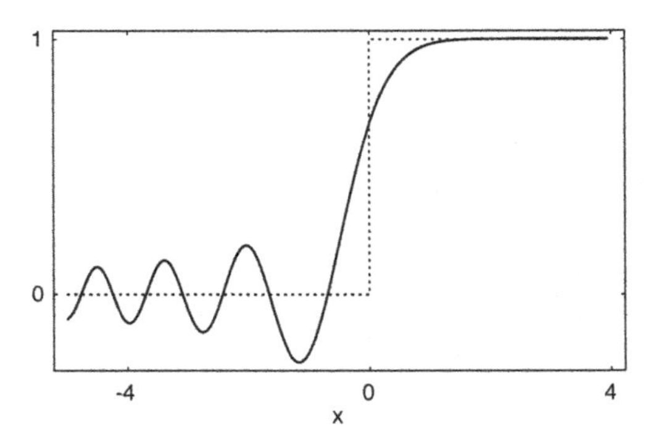

Fig. 4.2 Airy transform of the step function $\theta(x)$.

The transformed function oscillates for $x < 0$ and goes exponentially to 1 for $x > 0$ (Fig. (4.2)), as can be seen from the following asymptotic expansions

$$\varphi_\alpha(x) \approx 1 - \frac{1}{2\sqrt{\pi}} \left(\frac{x}{\alpha}\right)^{-3/4} \mathrm{e}^{-\frac{2}{3}\left(\frac{x}{\alpha}\right)^{3/2}},$$

when $x \to +\infty$, and

$$\varphi_\alpha(x) \approx \frac{1}{\sqrt{\pi}} \left(\frac{x}{\alpha}\right)^{-3/4} \cos\left[\frac{2}{3}\left(\frac{x}{\alpha}\right)^{3/2} + \frac{\pi}{4}\right],$$

when $x \to -\infty$.

We can also give an important result allowing the transform of many functions to be calculated analytically:

Proposition 4.7. *If the Airy transform of a function f is φ_α, then the Airy transform of xf is*

$$\mathcal{A}_\alpha[xf(x)] = x\varphi_\alpha - \alpha^3\varphi_\alpha''. \tag{4.44}$$

In fact the differentiation rule implies

$$\varphi_\alpha'' = f'' * \omega_\alpha = f * \omega_\alpha'',$$

but ω_α is a solution of the Airy equation $\omega_\alpha'' - \frac{x}{\alpha^3}\omega_\alpha = 0$. So we readily have:

$$\varphi_\alpha'' = \frac{1}{\alpha^3} f * (x\omega_\alpha) = \frac{1}{\alpha^3}(x(f * \omega_\alpha) - (xf) * \omega_\alpha),$$

which proves the result.

As we have seen, the Airy transform of 1 is 1, the transform of x is x,[1] and the Airy transform of x^2 is x^2.

This chain of transforms may be continued in order to form a family of polynomials: the *Airy polynomials*: $Pi_n(x)$. Some properties of these polynomials are given in the next section. Airy polynomials may be used to calculate the Airy transform of functions that can be written as a power series

$$f(x) = \sum_{n=0}^{\infty} c_n x^n. \tag{4.45}$$

The result in the transformation is a power series

$$\varphi_\alpha(x) = \sum_{n=0}^{\infty} \alpha^n c_n Pi_n\left(\frac{x}{\alpha}\right). \tag{4.46}$$

The preceding Proposition 4.7 can also be used to calculate the Airy transform of a function f, knowing the transform of xf. We have just to solve an inhomogeneous differential equation

$$\varphi'' - \frac{x}{\alpha^3}\varphi = \psi.$$

For example, we know that the Airy transform of 1 is 1. We can therefore calculate the transform of $f(x) = 1/x$. So that $\psi = \mathcal{A}_\alpha\left[x\frac{1}{x}\right] = 1$), and:

$$\varphi_\alpha(y) = \frac{1}{\alpha}\wp\int_{-\infty}^{+\infty} \frac{1}{x} Ai\left(\frac{y-x}{\alpha}\right) \mathrm{d}x = \frac{\pi}{\alpha} Gi\left[\frac{y}{\alpha}\right],$$

where $\wp$ is the Cauchy principal value and Gi is the inhomogeneous function (cf. §2.3.1) and $\alpha > 0$. The latter relation is simply formula (4.4).

Another useful result to calculate Airy transforms consists of a generalisation of the Plancherel–Parseval rule:

[1]To prove this result, and the one given by Eq. (4.46), we should introduce a converging factor (see §4.2.4.2).

Proposition 4.8. *If the Airy transform of the function* $f(x)$ *is* $\varphi_\alpha(x)$, *then these two functions have the same autocorrelation function*

$$F(x) = \int_{-\infty}^{+\infty} f(y)f(x+y)\mathrm{d}y = \int_{-\infty}^{+\infty} \varphi_\alpha(u)\varphi_\alpha(x+u)\mathrm{d}u. \tag{4.47}$$

The proof stems from the substitution of the value of φ_α into the right member of Eq. (4.47),

$$F(x) = \frac{1}{\alpha^2}\int_{\mathbb{R}^2} f(y)f(y')\mathrm{d}y\mathrm{d}y' \int_{-\infty}^{+\infty} Ai\left(\frac{u-y}{\alpha}\right) Ai\left(\frac{x+u-y}{\alpha}\right)\mathrm{d}u.$$

The formula (3.111) then leads to the required result.

If $\hat{F}(\xi)$ is the Fourier transform of $F(x)$, we can see that Eq. (4.47) may be reversed according to

$$f(x) = \frac{1}{2\pi}\int_{-\infty}^{+\infty} \sqrt{\hat{F}(\xi)}\mathrm{e}^{\mathrm{i}(\alpha^3\xi^3/3+\xi x)}\mathrm{d}\xi. \tag{4.48}$$

This relation represents an Airy transform in terms of the Fourier variable ξ (cf. Eq. (4.32)).[2] However, the inversion can also be performed by putting $\alpha = 0$. These results are easily checked in the particular case where f is a Gaussian or an Airy function. This inversion is an example of an ill-posed inverse problem.

We close this paragraph with a result that will be used in §7.5. In the case where the scaling parameter depends on another parameter: $\alpha = \alpha(t)$, we have the following lemma:

Proposition 4.9. *The Airy transform of the partial derivative with respect to the parameter* t *is given by*

$$\mathcal{A}_\alpha\left[\frac{\partial f}{\partial t}\right] = \frac{\partial \varphi_\alpha}{\partial t} + \dot{\alpha}\alpha^2\frac{\partial^3\varphi_\alpha}{\partial x^3} \tag{4.49}$$

where $\mathcal{A}_\alpha[f] = \varphi_\alpha$ *and* $\dot{\alpha} = \frac{\mathrm{d}\alpha}{\mathrm{d}t}$.

The proof is given by taking the Fourier transform of both sides of the equality. We readily see the importance of such a result in the Airy transform of an evolution equation. In particular, the case: $\dot{\alpha}\alpha^2 = \mathrm{C^{te}}$ was studied by Widder [Widder (1979)]. We shall employ this lemma in §7.5.1.1.

[2]This result is a particular case of a more general result of Wiener, see for instance Davis (1962), page 434.

Table 4.1 Some Airy transforms

$f(x)$	$\varphi_\alpha(y) = \frac{1}{\alpha}\int_{-\infty}^{+\infty} f(x)Ai\left(\frac{y-x}{\alpha}\right)\mathrm{d}x,\ \alpha > 0$
$f(x+k)$	$\varphi_\alpha(y+k)$
$f(kx)$	$\varphi_{\alpha k}(ky)$
$f'(x)$	$\varphi'_\alpha(y)$
$xf(x)$	$y\varphi_\alpha(y) - \alpha^3\varphi''_\alpha(y)$
1	1
x	y
x^2	y^2
x^n	$\alpha^n Pi_n\left(\frac{y}{\alpha}\right)$
e^{ikx}	$\exp\left[\mathrm{i}\left(ky + \alpha^3k^3/3\right)\right]$
$\delta(x)$	$\frac{1}{\alpha}Ai\left(\frac{y}{\alpha}\right)$
$\mathrm{sgn}(x)$	$1 - 2\int_{y/\alpha}^{+\infty} Ai(u)\mathrm{d}u$
$\theta(x)$	$1 - \int_{y/\alpha}^{+\infty} Ai(u)\mathrm{d}u$
$\frac{1}{x}$	$\frac{\pi}{\alpha}Gi\left[\frac{y}{\alpha}\right]$
$\frac{1}{\sqrt{\lvert x\rvert}}$	$\frac{2^{2/3}\pi}{\alpha^{1/2}}\left[Ai^2\left(\frac{y}{2^{2/3}\alpha}\right) + Ai\left(\frac{y}{2^{2/3}\alpha}\right)Bi\left(\frac{y}{2^{2/3}\alpha}\right)\right]$
e^{-x^2}	$\frac{\sqrt{\pi}}{\alpha}\exp\left[\frac{1}{4\alpha^3}\left(y + \frac{1}{24\alpha^3}\right)\right]Ai\left(\frac{y}{\alpha} + \frac{1}{16\alpha^4}\right)$
$Ai(x)$	$\left(\alpha^3+1\right)^{-1/3}Ai\left[\left(\alpha^3+1\right)^{-1/3}y\right]$
$Gi(x)$	$\left(\alpha^3+1\right)^{-1/3}Gi\left[\left(\alpha^3+1\right)^{-1/3}y\right]$
$Ai^2(x)$	$\left(4\alpha^3+1\right)^{-1/6}Ai^2\left[\left(4\alpha^3+1\right)^{-1/3}y\right]$
$Ai(x)Bi(x)$	$\left(4\alpha^3+1\right)^{-1/6}Ai\left[\left(4\alpha^3+1\right)^{-1/3}y\right]Bi\left[\left(4\alpha^3+1\right)^{-1/3}y\right]$

4.2.3 *Airy polynomials*

We define an Airy polynomial of degree n, $Pi_n(x)$, using the Airy transform of x^n (this is in fact a convolution product of two distributions)

$$Pi_n(x) = \int_{-\infty}^{+\infty} y^n Ai(x-y)\mathrm{d}y. \tag{4.50}$$

Taking the second derivative of this formula, we obtain directly the recurrence formula

$$Pi_{n+1}(x) = xPi_n(x) - Pi_n''(x). \tag{4.51}$$

Taking the derivative of this expression once again, we obtain the following third order differential equation satisfied by Airy polynomials

$$Pi_n'''(x) - xPi_n'(x) + nPi_n(x) = 0. \tag{4.52}$$

This equation has to be compared with formulae (2.104) and (2.148).[3] From the last two equations, we deduce the recurrence relation

$$Pi_{n+1}(x) = xPi_n(x) - n(n-1)Pi_{n-2}(x). \tag{4.53}$$

There is also an addition theorem for Airy polynomials. From definition (4.50), we can write

$$\sum_{k=0}^{n} \binom{n}{k} Pi_{n-k}(x)Pi_k(y) = \sum_{k=0}^{n} \binom{n}{k} \iint_{\mathbb{R}^2} (x')^{n-k}(y')^k Ai(x-x') \times Ai(y-y')\mathrm{d}x'\mathrm{d}y'.$$

Then, from the binomial relation $\sum\limits_{k=0}^{n} \binom{n}{k} x^{n-k}y^k = (x+y)^n$, this gives

$$\sum_{k=0}^{n} \binom{n}{k} Pi_{n-k}(x)Pi_k(y) = \iint_{\mathbb{R}^2} (x'+y')^n Ai(x-x')Ai(y-y')\mathrm{d}x'\mathrm{d}y'.$$

A change of variables and the use of formula (3.111) gives the addition theorem of these polynomials

$$\sum_{k=0}^{n} \binom{n}{k} Pi_{n-k}(x)Pi_k(y) = 2^{n/3} Pi_n\left(\frac{x+y}{2^{1/3}}\right). \tag{4.54}$$

Other properties of Airy polynomials can be mentioned: for instance their values at the origin (cf. formula (3.87))

$$\begin{cases} Pi_{3n}(0) = (-1)^n \frac{(3n)!}{3^n n!} \\ Pi_{3n+1}(0) = Pi_{3n+2}(0) = 0, \end{cases}$$

[3] We shall return to this kind of equation in §6.2.

and a generating function

$$e^{-t^3/3+xt} = \sum_{n=0}^{\infty} \frac{t^n}{n!} Pi_n(x). \tag{4.55}$$

Airy polynomials have similar properties to those of Hermite $He_n(x)$ [Abramowitz & Stegun (1965)]. As a matter of fact, Airy polynomials belong (like Hermite $He_n(x)$) to a family of polynomials — somewhat related to the so called Appell set [Erdélyi *et al.* (1981)] — sharing, in particular, the property

$$\frac{d}{dx} P_n(x) = n\, P_{n-1}(x).$$

Table (4.1) sums up the Airy transforms of some functions.

4.2.4 *A particular case: correlation Airy transform*

We now turn to a transform which can be deduced from the general case. If, in place of a convolution, we take the cross-correlation of a function by an Airy function, we can define

$$\widetilde{f}(x) = [f \star Ai](x) = \int_{-\infty}^{+\infty} f(y)\, Ai(x+y)\, dy = \mathfrak{A}[f](x), \tag{4.56}$$

where the integral is assumed to be convergent. We therefore call this transform the "Correlation Airy Transform (CAT)". It is a particular case of Airy transform (which may be seen as a convolution) sharing several noteworthy properties with the former one.

The connection with the general transform is given by

$$\mathfrak{A}[f](x) = \mathcal{A}_{-1}[f](-x).$$

In this form, this transform was first introduced by Vu Kim Tuan (1998) in connection with the Paley–Wiener theorem, and independently by Basor & Widom (1999) in connection with the Airy kernel. In both papers, it is shown that this transform may be seen as a composition of two inverse Fourier transforms

$$\mathfrak{A}[f] = (\mathcal{F}^{-1} e^{it^3/3} \mathcal{F}^{-1})[f].$$

4.2.4.1 *Properties of the correlation Airy transform*

We now give some properties of the CAT. All the integrals below are assumed to be convergent, or at least semi-convergent. The first property we are going to derive is the most important one. Contrary to the general Airy transform, we have

Proposition 4.10. *The* CAT *is an involutive transform i.e.* $\mathfrak{A}$ *is a unitary operator.*

$$\mathfrak{A}\mathfrak{A}[f](x) = f(x). \tag{4.57}$$

Proof:[4]

$$\mathfrak{A}\mathfrak{A}[f](x) = \mathfrak{A}\widetilde{f}(x) = \int_{-\infty}^{+\infty}\int_{-\infty}^{+\infty} f(u)\, Ai(x+y)Ai(u+y)\,\mathrm{d}y\,\mathrm{d}u.$$

But from Eq. (3.111), we have $\int_{-\infty}^{+\infty} Ai(x+y)Ai(u+y)\,\mathrm{d}y = \delta(u-x)$. □

The result (4.57) was formerly shown by Vu Kim Tuan (1998) and Basor and Widom (1999).

Proposition 4.11. *The direction of a translation is reverted in a* CAT*:*

$$\mathfrak{A}[f](x+a) = \widetilde{f}(x-a). \tag{4.58}$$

The proof is given by a simple change of variable.

Proposition 4.12. *The transform of the derivative is the opposite of the derivative of the transform:*

$$\mathfrak{A}[f'](x) = -\frac{\mathrm{d}}{\mathrm{d}x}\Big(\mathfrak{A}[f](x)\Big) = -(\widetilde{f})'(x). \tag{4.59}$$

The proof is given by an integration by parts.

Proposition 4.13. *The transform of* $xf(x)$ *is given by*

$$\mathfrak{A}[xf(x)] = (\widetilde{f})''(x) - x\widetilde{f}(x). \tag{4.60}$$

Proof:

$$\begin{aligned}\mathfrak{A}[xf(x)] &= \int_{-\infty}^{+\infty} Ai(x+u)\, uf(u)\,\mathrm{d}u \\ &= \int_{-\infty}^{+\infty} (u+x)\, Ai(x+u)\, f(u)\mathrm{d}u - x\int_{-\infty}^{+\infty} Ai(x+u)f(u)\mathrm{d}u.\end{aligned}$$

[4]The □ means 'End of proof'

But from the Airy equation $Z'' = xZ$, we find

$$\int_{-\infty}^{+\infty} (u+x)\,Ai(x+u) f(u)\mathrm{d}u = \int_{-\infty}^{+\infty} Ai''(x+u) f(u)\mathrm{d}u.$$

Then two integrations by parts give the result. □

Proposition 4.14. *The* CAT *transforms a convolution product into a cross-correlation product:*

$$\mathfrak{A}[f * g] = f \star \widetilde{g} = g \star \widetilde{f}. \tag{4.61}$$

Proof: Subject to the convergence of the integrals

$$\begin{aligned}\mathfrak{A}[f * g](x) &= \int_{-\infty}^{+\infty}\int_{-\infty}^{+\infty} f(t) g(u-t) Ai(x+u)\,\mathrm{d}u\mathrm{d}t \\ &= \int_{-\infty}^{+\infty} f(t)\,\widetilde{g}(x+t)\,\mathrm{d}t = f \star \widetilde{g}(x). \;\square\end{aligned}$$

Proposition 4.15. *The* CAT *transforms a cross-correlation product into a convolution product:*

$$\mathfrak{A}[f \star g] = f * \widetilde{g}. \tag{4.62}$$

The proof is similar to the one above.

Proposition 4.16. *A function and its transform have the same auto-correlation function.*

$$f \star f = \widetilde{f} \star \widetilde{f}. \tag{4.63}$$

Proof: Subject to the convergence of the integrals, we have formally

$$\begin{aligned}\int_{-\infty}^{+\infty} \widetilde{f}(t)\widetilde{g}(t+x)dt &= \iiint_{-\infty}^{+\infty} Ai(u+t)Ai(v+t+x) f(u)g(v)\,\mathrm{d}u\,\mathrm{d}v\,\mathrm{d}t \\ &= \iint_{-\infty}^{+\infty} f(u)g(v)\delta(x+v-u)\,\mathrm{d}u\,\mathrm{d}v \\ &= \int_{-\infty}^{+\infty} f(u+x)g(u)\,\mathrm{d}u.\end{aligned}$$

Then $\widetilde{f} \star \widetilde{g} = g \star f$, which gives the result when $f = g$. □

Proposition 4.17. *(Plancherel–Parseval theorem) Putting $x = 0$ in the preceding result, we find*

$$\int_{-\infty}^{+\infty} \widetilde{f}(u)\widetilde{g}(u)du = \int_{-\infty}^{+\infty} f(u)g(u)\,\mathrm{d}u. \tag{4.64}$$

4.2.4.2 *Some examples*

The first examples are those of polynomials. In this case, the convergence is not ensured. We therefore introduce a converging factor and then have to evaluate

$$I_n = \lim_{\varepsilon \to 0} \int_{-\infty}^{+\infty} \mathrm{e}^{\varepsilon t} Ai(x+t)\, t^n \,\mathrm{d}t. \tag{4.65}$$

In the case $n = 0$, we have, from Eq. (3.99): $\int_{-\infty}^{+\infty} \mathrm{e}^{\varepsilon t} Ai(t)\,\mathrm{d}t = \mathrm{e}^{\varepsilon^3/3}$. Then

$$I_0 = \lim_{\varepsilon \to 0} \mathrm{e}^{\varepsilon^3/3 - \varepsilon x} = 1, \tag{4.66}$$

which justifies the relation (4.30). We now give the cases $n = 1, 2, 3$, which can be obtained from the derivatives

$$I_1 = \lim_{\varepsilon \to 0} \frac{\mathrm{d}}{\mathrm{d}\varepsilon} \left(\mathrm{e}^{\varepsilon^3/3 - \varepsilon x} \right) = -x, \tag{4.67}$$

$$I_2 = \lim_{\varepsilon \to 0} \frac{\mathrm{d}^2}{\mathrm{d}\varepsilon^2} \left(\mathrm{e}^{\varepsilon^3/3 - \varepsilon x} \right) = x^2, \tag{4.68}$$

$$I_3 = \lim_{\varepsilon \to 0} \frac{\mathrm{d}^3}{\mathrm{d}\varepsilon^3} \left(\mathrm{e}^{\varepsilon^3/3 - \varepsilon x} \right) = 2 - x^3. \tag{4.69}$$

These integrals are clearly related to the Airy polynomials.

The second example is also singular, and will be treated as a Cauchy principal value integral: $\wp \int_{-\infty}^{+\infty} Ai(t+x)\, \frac{1}{t}\,\mathrm{d}t$. In fact, we have seen that

$$\wp \int_{-\infty}^{+\infty} \frac{Ai(u)}{u-x}\,\mathrm{d}u = \wp \int_{-\infty}^{+\infty} Ai(t+x)\, \frac{1}{t}\,\mathrm{d}t = -\pi\, Gi(x). \tag{4.70}$$

This result is generalised as follows. In his paper, Moyer (1973) derived the Green's function of a particle in a uniform electric field. His result may alternatively be written (see §4.1.2):

$$\begin{aligned} \frac{1}{2\pi^{3/2}} \int_0^{+\infty} \exp\left\{ \mathrm{i} \left[\frac{t^3}{12} + \frac{u+v}{2} t - \frac{(u-v)^2}{4t} + \frac{\pi}{4} \right] \right\} \frac{\mathrm{d}t}{\sqrt{t}} &= \\ = \begin{cases} Ai(u)[Ai(v) + \mathrm{i}Bi(v)] \text{ if } u \geq v, \\ \\ Ai(v)[Ai(u) + \mathrm{i}Bi(u)] \text{ if } u \leq v. \end{cases} \end{aligned} \tag{4.71}$$

This integral (Eq. (4.71)) may be seen as a Fourier transform. Then, if we put $u = x + w$ and $v = y + w$, using the causal properties of Fourier transforms, the real and imaginary parts of this result can be written as a Hilbert transform

$$\wp \int_{-\infty}^{+\infty} \frac{Ai(t+x)Ai(t+y)}{t-w}\,\mathrm{d}t = -\pi \begin{cases} Ai(x+w)Bi(y+w) & \text{if } x \geq y, \\ Ai(y+w)Bi(x+w) & \text{if } x \leq y. \end{cases} \tag{4.72}$$

This result can be seen as the CAT of $Ai(z+y)/(z-w)$. From this result, we also can give the CAT of the derivative $Ai'(z+y)/(z-w)$ and, by inverting the Hilbert transform, the CAT of $Bi(z+y)/(z-w)$.

We now give some other transforms (without demonstration) which are less singular than the previous ones. First, we have for the transform of a normalized Gaussian ($u > 0$)

$$\frac{1}{2\sqrt{\pi u}} \int_{-\infty}^{+\infty} \mathrm{e}^{-\frac{t^2}{4u}}\, Ai(t+x)\,\mathrm{d}t = \mathrm{e}^{2u^3/3+ux}\, Ai(x+u^2). \tag{4.73}$$

Then we give the transform of $\mathrm{e}^{ux} Ai(ax)$, $u > 0$

$$\int_{-\infty}^{+\infty} \mathrm{e}^{ut} Ai(at)\, Ai(x+t)\,\mathrm{d}t$$
$$= \begin{cases} \frac{1}{2\sqrt{\pi u}} \exp\left(\frac{u^3}{12} - \frac{x^2}{4u} - \frac{ux}{2}\right), & \text{if } a = 1 \\ \frac{1}{|a^3-1|^{1/3}} \exp\left[\frac{u^3}{3}\frac{a^3+1}{(a^3-1)^2} + \frac{ux}{a^3-1}\right] & \\ Ai\left[\frac{a\,\mathrm{sgn}(a^3-1)}{|a^3-1|^{1/3}}\left(x + \frac{u^2}{a^3-1}\right)\right], & \text{if } a \neq 1. \end{cases} \tag{4.74}$$

which justifies the semi-convergent integral Eq. (3.111) in the limit $u \to 0$.

Next, two transforms involving products of Airy functions

$$\int_{-\infty}^{+\infty} Ai(a-t)Ai(b-t)\, Ai(x+t)\,\mathrm{d}t$$
$$= 5^{-1/6} Ai\left[5^{-1/3}\left(x + \frac{a}{\eta} - \eta b\right)\right] Ai\left[5^{-1/3}\left(x + \frac{b}{\eta} - \eta a\right)\right], \tag{4.75}$$

where $\eta = \frac{\sqrt{5}-1}{2}$ is the Golden mean, and

$$\int_{-\infty}^{+\infty} Ai(-t)Ai(t)\, Ai(x+t)\,\mathrm{d}t = 5^{-1/6} Ai\left[5^{-1/3}\frac{x}{\eta}\right] Ai\left[-5^{-1/3}x\eta\right]. \tag{4.76}$$

Now, we consider the following integral: $\int_{-\infty}^{+\infty} Ai^2(at)\, Ai(x+t)\,\mathrm{d}t$, which is convergent when $a < 0$. However, the result may be extended to the semi-convergent case.

$$\int_{-\infty}^{+\infty} Ai^2(at)\, Ai(x+t)\,\mathrm{d}t =$$
$$\begin{cases} \frac{1}{(1-4a^3)^{1/6}}\, Ai^2\left[\frac{-a}{(1-4a^3)^{1/3}}x\right], & \text{if } a < 2^{-2/3} \\ \frac{1}{(4a^3-1)^{1/6}}\, Ai\left[\frac{a}{(4a^3-1)^{1/3}}x\right] Bi\left[\frac{a}{(4a^3-1)^{1/3}}x\right], & \text{if } a > 2^{-2/3}. \end{cases} \tag{4.77}$$

In order to prove, at least formally, this result, we consider the following integral

$$J = \int_{-\infty}^{+\infty} \left[Ai^2(at) + \mathrm{i}Ai(at)Bi(at)\right] Ai(x+t)\,\mathrm{d}t. \tag{4.78}$$

Using the integral representations

$$Ai(x+t) = \frac{1}{2\pi}\int_{-\infty}^{+\infty} \exp\left\{\mathrm{i}\left[\frac{v^3}{3} + v(x+t)\right]\right\}\mathrm{d}v, \tag{4.79}$$

and

$$Ai^2(at) + \mathrm{i}Ai(at)Bi(at) = \frac{1}{2\pi^{3/2}}\int_0^{+\infty} \exp\left\{\mathrm{i}\left[\frac{u^3}{12} + atu + \frac{\pi}{4}\right]\right\}\frac{\mathrm{d}u}{\sqrt{u}}, \tag{4.80}$$

where $u > 0$ (see Eq. (4.71)), the integral (Eq. (4.78)) now reads

$$\frac{1}{4\pi^{5/2}}\int_0^{+\infty}\int_{-\infty}^{+\infty}\int_{-\infty}^{+\infty} \exp\left\{\mathrm{i}\left[\frac{4v^3+u^3}{12} + vx + \frac{\pi}{4}\right]\right\} \mathrm{e}^{\mathrm{i}(au+v)t}\,\mathrm{d}v\,\frac{\mathrm{d}u}{\sqrt{u}}\,\mathrm{d}t. \tag{4.81}$$

An integration on t leads to

$$\frac{1}{2\pi^{3/2}}\int_0^{+\infty}\int_{-\infty}^{+\infty} \exp\left\{\mathrm{i}\left[\frac{4v^3+u^3}{12} + vx + \frac{\pi}{4}\right]\right\}\delta(au+v)\,\mathrm{d}v\,\frac{\mathrm{d}u}{\sqrt{u}}. \tag{4.82}$$

Then we have

$$J = \frac{1}{2\pi^{3/2}}\int_0^{+\infty} \exp\left\{\mathrm{i}\left[\frac{(1-4a^3)u^3}{12} - aux + \frac{\pi}{4}\right]\right\}\frac{\mathrm{d}u}{\sqrt{u}}. \tag{4.83}$$

If $a < 2^{-2/3}$, we make the change of variable $w = u(1-4a^3)^{1/3}$ and, by separating the real and imaginary parts, we have

$$\int_{-\infty}^{+\infty} Ai^2(at)\, Ai(x+t)\,\mathrm{d}t = \frac{1}{(1-4a^3)^{1/6}}\, Ai^2\left[\frac{-ax}{(1-4a^3)^{1/3}}\right]. \tag{4.84}$$

But if $a > 2^{-2/3}$, we have to take the complex conjugate of J and, multiply the integral by i, leading to

$$\mathrm{i}J^* = \frac{\mathrm{e}^{\mathrm{i}\pi/2}}{2\pi^{3/2}}\int_0^{+\infty} \exp\left\{\mathrm{i}\left[\frac{(4a^3-1)u^3}{12} + aux - \frac{\pi}{4}\right]\right\}\frac{\mathrm{d}u}{\sqrt{u}}. \tag{4.85}$$

Making the change of variable $w = u(4a^3 - 1)^{1/3}$, and again separating the real and imaginary parts, we have

$$\int_{-\infty}^{+\infty} Ai^2(at)Ai(x+t)\,\mathrm{d}t = \frac{1}{(4a^3-1)^{1/6}}\, Ai\left[\frac{ax}{(4a^3-1)^{1/3}}\right] Bi\left[\frac{ax}{(4a^3-1)^{1/3}}\right]. \tag{4.86}$$

For the case $a = 2^{-2/3}$, we use a result from Aspnes (1966) (see Eq. (3.89))

$$\int_0^{\infty} Ai(x+t)\frac{\mathrm{d}t}{\sqrt{t}} = 2^{2/3}\pi Ai^2(2^{-2/3}x), \tag{4.87}$$

which is the CAT of $\theta(x)/\sqrt{x}$. Then by inverse transform, we have

$$\int_{-\infty}^{+\infty} Ai^2(2^{-2/3}t)Ai(x+t)\,\mathrm{d}t = \frac{1}{2^{2/3}\pi}\frac{1}{\sqrt{x}}\theta(x), \tag{4.88}$$

where θ is the step function. Conversely, we also have

$$\int_{-\infty}^{+\infty} Ai(at)Bi(at)\,Ai(x+t)\,\mathrm{d}t =$$

$$\begin{cases} \frac{1}{(1-4a^3)^{1/6}}\, Ai\left[\frac{-a}{(1-4a^3)^{1/3}}x\right] Bi\left[\frac{-a}{(1-4a^3)^{1/3}}x\right], & \text{if } a < 2^{-2/3}; \\ \frac{1}{(4a^3-1)^{1/6}}\, Ai^2\left[\frac{a}{(4a^3-1)^{1/3}}x\right], & \text{if } a > 2^{-2/3}. \end{cases} \tag{4.89}$$

For the particular case $a = 2^{-2/3}$, we have again from Aspnes (1966)

$$\int_0^{\infty} Ai(x-t)\frac{\mathrm{d}t}{\sqrt{t}} = \int_{-\infty}^{0} Ai(x+t)\frac{\mathrm{d}t}{\sqrt{-t}} = 2^{2/3}\pi Ai(2^{-2/3}x)Bi(2^{-2/3}x),$$

which gives, by inverse transform,

$$\int_{-\infty}^{+\infty} Ai(2^{-2/3}t)Bi(2^{-2/3}t)Ai(x+t)\,\mathrm{d}t = \frac{1}{2^{2/3}\pi}\frac{1}{\sqrt{-x}}\theta(-x). \tag{4.90}$$

The particular case $a = 2^{-1/3}$ gives a self-reciprocal transform (see the next section)

$$\int_{-\infty}^{+\infty} \left[Ai^2(2^{-1/3}t) + Ai(2^{-1/3}t)Bi(2^{-1/3}t)\right] Ai(x+t)\,\mathrm{d}t =$$

$$Ai(2^{-1/3}t)Bi(2^{-1/3}t) + Ai^2(2^{-1/3}t), \tag{4.91}$$

a result which can be found from Eqs. (4.86) and (4.89). If, in Eqs. (4.77) and (4.89), we put $a = 1$, $x = 0$, we find

$$\int_{-\infty}^{+\infty} Ai^3(t)\,\mathrm{d}t = 3^{-1/6}Ai(0)Bi(0) = \frac{\Gamma^2(1/3)}{4\pi^2} = \sqrt{3}\int_{-\infty}^{+\infty} Ai^2(t)\,Bi(t)\,\mathrm{d}t,$$

a noteworthy result already found by Reid (1997a) using another method (see also Varlamov (2008-a)). It should be noted that, in the case where $a < 2^{-2/3}$, the transform maps the open interval $-\infty < a < 0$ onto the open interval $0 < a < 2^{-2/3}$, and *vice versa*.

Naturally, from Eqs. (4.58), (4.59) and (4.60), many other transforms could be found.

4.2.4.3 Self-reciprocal transforms

Let us consider the following Fredholm integral equation of the second kind

$$\varphi(x) - \lambda \int_{-\infty}^{+\infty} Ai(x+u)\,\varphi(u)\,\mathrm{d}u = f(x), \tag{4.92}$$

where φ is an unknown function. Note that the kernel of this equation is not square integrable. It can also be written : $(1-\lambda\,\mathfrak{A})\varphi(x) = f(x)$. Making a CAT of this equation we find $\mathfrak{A}\varphi(x) - \lambda\,\varphi(x) = \mathfrak{A}f(x)$. Then, rearranging this last equation with Eq. (4.92), gives (with $\lambda^2 \neq 1$) the solution

$$\varphi(x) = \frac{1}{1-\lambda^2}\left(f(x) + \lambda \int_{-\infty}^{+\infty} Ai(x+u)\,f(u)\,du\right). \tag{4.93}$$

The solution we have found suggests that $\lambda = \pm 1$ are the characteristic numbers of the homogeneous Fredholm equation

$$\varphi(x) = \lambda \int_{-\infty}^{+\infty} Ai(x+u)\,\varphi(u)\,\mathrm{d}u. \tag{4.94}$$

Two propositions are found for this integral equation. Let us call the families of eigenfunctions $\mathcal{R}^+$ for the eigenvalue $+1$ and $\mathcal{R}^-$ for the eigenvalue -1.

Lemma 4.18. *The family $\mathcal{R}^-$ is obtained from the derivative of the functions of $\mathcal{R}^+$.*

Proof: Taking an eigenfunction $\varphi^+ \in \mathcal{R}^+$, we have from the property Eq. (4.59): $(\mathfrak{A}\varphi^+)' = -(\varphi^+)'$, then the derivative of φ^+ belongs to $\mathcal{R}^-$. □

Lemma 4.19. *The functions of $\mathcal{R}^+$ are orthogonal to those of $\mathcal{R}^-$.*

Proof: According to the Pancherel–Parseval theorem Eq. (4.64), we have

$$\int_{-\infty}^{+\infty} \varphi^+\varphi^-\mathrm{d}x = \int_{-\infty}^{+\infty} \mathfrak{A}\varphi^+\,\mathfrak{A}\varphi^-\mathrm{d}x = -\int_{-\infty}^{+\infty} \varphi^+\varphi^-\mathrm{d}x = 0. \quad \square$$

We now derive some of the eigenfunctions of Eq. (4.94). First, we consider the case $\lambda = 1$. Clearly $\varphi(x) = 1$ (or a constant) is an eigenfunction (see Eq. (4.66)), but x^2 also thanks to Eq. (4.68), and so forth. Furthermore, there are other eigenfunctions for the characteristic number $\lambda = 1$. Using Eq. (3.111) gives the self-reciprocal transform

$$\int_{-\infty}^{+\infty} Ai(x+t)Ai(2^{1/3}t)\,\mathrm{d}t = Ai(2^{1/3}x). \tag{4.95}$$

Thus, in order to find other eigenfunctions, we can take the second derivative of the preceding result (Eq. (4.95)) to obtain: $xAi(2^{1/3}x)$, and for $\lambda = -1$, $Ai'(2^{1/3}x)$, ... , thereby obtaining two other infinite sets of eigenfunctions.

Lastly, we give a parametrized family generated by

$$\varphi_0^+(x,u) = \mathrm{e}^{ux} Ai[2^{1/3}(x+u^2/2)],\ u>0,$$

from which we can derive an orthogonality example

$$\int_{-\infty}^{+\infty} \mathrm{e}^{ux} Ai[2^{1/3}(x+u^2/2)] Ai'(2^{1/3}x)\,\mathrm{d}x = 0,\ \forall u > 0.$$

4.2.4.4 *Application to the Airy kernel*

The Airy kernel has appeared in several different studies such as for instance, as the kernel of a linear integral equation associated to the (nonlinear) Painlevé PII equation [Clarkson, (2003)], or in Fredholm determinants of integral operators appearing as spectral distribution functions for random matrices [Tracy & Widom (1994)]. Recently the Airy kernel appeared as the matrix element of the projection operator for a discrete variable representation basis [Littlejohn & Cargo (2002)], in view of applications to chemical physics. More recently, we can also mention the work of Levin & Lubinsky (2009). The Airy kernel is defined by

$$N(x,y;z) = \frac{1}{x-y}\left[Ai(x+z)Ai'(y+z) - Ai'(x+z)Ai(y+z)\right]. \quad (4.96)$$

We first recall the result given above (Eq. (4.72))

$$\wp \int_{-\infty}^{+\infty} \frac{Ai(t+u')Ai(t+v')}{t-w}\,\mathrm{d}t$$

$$= -\pi \begin{cases} Ai(u'+w)Bi(v'+w) \text{ if } u' \geq v', \\ \\ Ai(v'+w)Bi(u'+w) \text{ if } u' \leq v'. \end{cases}$$

This result may also be considered as the CAT of $Ai(t+u')/(t-w)$.

We now make a CAT of the Airy kernel, $\int_{-\infty}^{+\infty} Ai(x+x')N(x,y;z)\,\mathrm{d}x$. First assuming $x' \geq z$, we can express the first term inside the square brackets in Eq. (4.96) as

$$Ai'(y+z)\int_{-\infty}^{+\infty} \frac{Ai(x+x')Ai(x+z)}{x-y}\,\mathrm{d}x = -\pi Ai'(y+z)Bi(z+y)Ai(x'+y).$$

For the second term inside the square brackets in Eq. (4.96), using the derivative of Eq. (4.72), we have

$$Ai(y+z)\int_{-\infty}^{+\infty}\frac{Ai(x+x')Ai'(x+z)}{x-y}\,\mathrm{d}x = \pi Ai(y+z)Bi'(z+y)Ai(x'+y).$$

The sum of these last two equations is rather simple, for we recognize the Wronskian of Airy functions i.e. $1/\pi$. Now in the case where $x' \leq z$, we can express the first term inside the square brackets in Eq. (4.96) as

$$Ai'(y+z)\int_{-\infty}^{+\infty}\frac{Ai(x+x')Ai(x+z)}{x-y}\,\mathrm{d}x = -\pi Ai'(y+z)Bi(x'+y)Ai(z+y).$$

For the second term, we have

$$Ai(y+z)\int_{-\infty}^{+\infty}\frac{Ai(x'+x)Ai'(x+z)}{x-y}\,\mathrm{d}x = \pi Ai'(y+z)Bi(x'+y)Ai(z+y).$$

The sum of these last two equations cancels. Finally, we obtain the result

$$\int_{-\infty}^{+\infty} Ai(x+x')N(x,y;z)\,\mathrm{d}x = Ai(x'+y)\theta(x'-z), \tag{4.97}$$

where θ is the Heaviside step function. Note however, that this result may be obtained more directly from another form of the Airy kernel, since the inverse transform is

$$N(x,y;z) = \int_{z}^{+\infty} Ai(x+x')Ai(y+x')\,\mathrm{d}x', \tag{4.98}$$

a result which was already known [Ablowitz & Segur (1981)]. We simply have

$$\int_{-\infty}^{+\infty} Ai(x+x')\int_{z}^{+\infty} Ai(x+u)Ai(y+u)\,\mathrm{d}x\,\mathrm{d}u = Ai(x'+y)\theta(x'-z),$$

by exchanging the two integrals.

4.2.4.5 *The* CAT *of Hermite functions*

Consider the Schrödinger equation for the harmonic oscillator (see for instance Landau & Lifschitz (1966))

$$\frac{\mathrm{d}^2\psi}{\mathrm{d}x^2} + \left(\epsilon_n - \frac{x^2}{4u^2}\right)\psi = 0, \tag{4.99}$$

where $\epsilon_n = (n+1/2)/u$, $n \in \mathbb{N}$, represents the eigenvalues of this quantum system and u is related to some physical parameters. The Hermite eigenfunctions are given by

$$\psi_n(x) = N_n \mathrm{e}^{-\frac{x^2}{4u}} H_n\left(\frac{x}{\sqrt{2u}}\right), \tag{4.100}$$

H_n being the Hermite polynomials and $N_n = \left(\frac{1}{2\pi u}\right)^{1/4} \frac{1}{\sqrt{2^n n!}}$ the normalisation constant.

The Hermite functions can be found recursively from the relation

$$\psi_n(x) = \frac{1}{2\sqrt{un}} \left(x - 2u\frac{\mathrm{d}}{\mathrm{d}x} \right) \psi_{n-1}(x). \tag{4.101}$$

Making a CAT of this recurrence relation yields, for the transformed functions,

$$\widetilde{\psi}_n(x) = \frac{1}{2\sqrt{un}} \left((\widetilde{\psi}_{n-1})''(x) + 2u(\widetilde{\psi}_{n-1})'(x) - x\widetilde{\psi}_{n-1}(x) \right), \tag{4.102}$$

by virtue of the properties Eqs. (4.59) and (4.60). The transformed ground-state wavefunction ($n = 0$) is given from the integral Eq. (4.73) as

$$\widetilde{\psi}_0(x) = (8\pi u)^{1/4}\, \mathrm{e}^{2u^3/3+ux}\, Ai(x+u^2), \tag{4.103}$$

and from the recurrence relation, we can derive the next transformed functions

$$\begin{aligned}
\widetilde{\psi}_1(x) &= M_1(u)\, \mathrm{e}^{ux} \left[uAi(x+u^2) + Ai'(x+u^2)\right] \\
\widetilde{\psi}_2(x) &= M_2(u)\, \mathrm{e}^{ux} \left[(4ux + 8u^3 + 1)Ai(x+u^2) + 8u^2 Ai'(x+u^2)\right] \\
&\vdots
\end{aligned}$$

where $M_n(u)$ is the normalisation constant.

Finally, thanks to the Plancherel–Perseval theorem, we have obtained a complete orthonormal basis $\{\widetilde{\psi}_n(x)\,|n \in \mathbb{N}\}$ of $\mathrm{L}_2(\mathbb{R})$

$$\int_{-\infty}^{+\infty} \widetilde{\psi}_n(x)\widetilde{\psi}_m(x)\mathrm{d}x = \int_{-\infty}^{+\infty} \psi_n(x)\psi_m(x)\mathrm{d}x = \delta_{n,m}, \tag{4.104}$$

where $\widetilde{\psi}_n(x)$ must satisfy the following fourth order eigenvalue differential equation, which is obtained from the CAT of Eq. (4.99):

$$\frac{\mathrm{d}^4\widetilde{\psi}_n(x)}{\mathrm{d}x^4} - (2x+4u^2)\frac{\mathrm{d}^2\widetilde{\psi}_n(x)}{\mathrm{d}x^2} - 2\frac{\mathrm{d}\widetilde{\psi}_n(x)}{\mathrm{d}x} + (x^2 - 4u^2\epsilon_n)\widetilde{\psi}_n(x) = 0, \tag{4.105}$$

where the eigenvalues $\epsilon_n = (n + 1/2)/u$, are evidently the same as for Eq. (4.99). This equation is readily self-adjoint.

Let us introduce the operator $b^\dagger = \frac{1}{\sqrt{2}}(\frac{\mathrm{d}^2}{\mathrm{d}x^2} + \frac{\mathrm{d}}{\mathrm{d}x} - x)$ and its adjoint $b = \frac{1}{\sqrt{2}}(\frac{\mathrm{d}^2}{\mathrm{d}x^2} - \frac{\mathrm{d}}{\mathrm{d}x} - x)$, where we have chosen the convenient value for the parameter $u = 1/2$. The operator $b^\dagger$ in fact comes from the recurrence relation (4.102). Note also that $\frac{1}{\sqrt{2}}(b + b^\dagger)$ is the Airy operator. The commutator of these two operators is easily found to be $[b, b^\dagger] = 1$, while the operator associated to the differential equation (4.105) is given by:

$$L = \frac{1}{2}\left(b\,b^\dagger + b^\dagger b\right) = b^\dagger b + \frac{1}{2} = L^\dagger,$$

so that Eq. (4.105) now reads $L\widetilde{\psi}_n = \epsilon_n\,\widetilde{\psi}_n$. Finally, the transformed equation has the same algebra of creation and annihilation (raising and lowering) operators as the harmonic oscillator. In particular, it shares the relationships $[b^\dagger b, b] = -b$ and $[b^\dagger b, b^\dagger] = b^\dagger$.

4.2.4.6 *Airy averaging*

Some years ago Englert and Schwinger (1984) introduced a method they called Airy averaging, in the context of an improvement of the Thomas–Fermi statistical model (see also Arrighini *et al.* (1999)). The definition of the Airy averaging of a function $f(x)$ is

$$\langle f(x)\rangle_{Ai} = \int_{-\infty}^{\infty} Ai(x) f(x)\,\mathrm{d}\,x. \tag{4.106}$$

In their paper Englert and Schwinger (1984) gave several properties of Airy averaging, such as

$$\langle x\rangle_{Ai} = 0, \quad \langle x^2\rangle_{Ai} = 0, \quad \langle x^3\rangle_{Ai} = 2$$
$$\langle x^{n+1}\rangle_{Ai} = n(n-1)\langle x^{n-2}\rangle_{Ai}$$
$$\langle x f(x)\rangle_{Ai} = \langle \frac{\mathrm{d}^2}{\mathrm{d}x^2} f(x)\rangle_{Ai}.$$

Clearly, from the definition and properties, Airy averaging is a particular case of the correlation Airy transform, for we have

$$\langle f(x)\rangle_{Ai} = \mathfrak{A}[f](0)\,. \tag{4.107}$$

In particular, we have $\langle x^{3n}\rangle_{Ai} = (-1)^n Pi_n(0)$, and all the other properties of Airy averaging can be deduced from those of the correlation Airy transform.

4.3 Other kinds of transformations

4.3.1 *Laplace transform of Airy functions*

In 1982 Davison & Glasser calculated the Laplace transform of the functions $Ai(-x)$ and $Bi(-x)$, $x > 0$, for their application in surface physics

(Schrödinger equation in a uniform electric field). In 1983, Leach calculated the Laplace transform of $Ai(\pm x)$ and $Bi(-x)$, $x > 0$, from their integral representation, for their application to magnetohydrodynamics. Exton (1985) and Wille (1986) calculated the Laplace transform of the Meijer function $G = G_{p,q}^{m,n}\left(z \middle| \begin{matrix} a_1, \cdots, a_p \\ b_1, \cdots, b_q \end{matrix}\right)$. We shall not detail the definition of this function (see Gradshteyn & Ryzhik (1965)), but are concerned with a particular case of G $G_{0,2}^{2,0}\left(x^3 \,|a, b\right) = 2x^{3(a+b)/2} K_{a-b}\left(2x^{3/2}\right)$, with $(a, b) = \left(\frac{1}{3}, 0\right)$, $\left(\frac{2}{3}, 0\right)$ or $\left(\frac{2}{3}, \frac{1}{3}\right)$, K_ν being the modified Bessel function (cf. §2.2.4). We therefore obtain the Laplace transforms of Airy functions:

$$I_{Ai_-}(p) = \int_0^\infty \mathrm{e}^{-px} Ai(-x)\mathrm{d}x = \frac{\mathrm{e}^a}{3}\left[2 - \frac{\gamma\left(\frac{2}{3}, a\right)}{\Gamma\left(\frac{2}{3}\right)} - \frac{\gamma\left(\frac{1}{3}, a\right)}{\Gamma\left(\frac{1}{3}\right)}\right] \tag{4.108}$$

$$I_{Bi_-}(p) = \int_0^\infty \mathrm{e}^{-px} Bi(-x)\mathrm{d}x = \frac{\mathrm{e}^a}{\sqrt{3}}\left[-\frac{\gamma\left(\frac{2}{3}, a\right)}{\Gamma\left(\frac{2}{3}\right)} + \frac{\gamma\left(\frac{1}{3}, a\right)}{\Gamma\left(\frac{1}{3}\right)}\right] \tag{4.109}$$

$$I_{Ai_+}(p) = \int_0^\infty \mathrm{e}^{-px} Ai(x)\mathrm{d}x = \frac{\mathrm{e}^{-a}}{3}\left[1 + \frac{\phi\left(\frac{2}{3}, a\right)}{\Gamma\left(\frac{2}{3}\right)} - \frac{\phi\left(\frac{1}{3}, a\right)}{\Gamma\left(\frac{1}{3}\right)}\right], \tag{4.110}$$

with $a = \frac{p^3}{3}$. $\Gamma(x)$ is the gamma function, $\gamma(\alpha, x)$ is the incomplete gamma function, $\gamma(\alpha, x) = \int_0^x \mathrm{e}^{-u} u^{\alpha-1}\mathrm{d}u$, and $\phi(\alpha, x) = \int_0^x \mathrm{e}^{u} u^{\alpha-1}\mathrm{d}u$. We can see that it is possible to express these transforms in terms of the confluent hypergeometric function F, thanks to the relations

$$\gamma(\alpha, x) = \frac{x^\alpha}{\alpha}\mathrm{e}^{-x}\, F\left(1, 1+\alpha; x\right) \quad \text{and} \quad \phi(\alpha, x) = \frac{x^{1-\alpha}}{1-\alpha}\, F\left(1-\alpha, 2-\alpha; x\right).$$

4.3.2 *Mellin transform of Airy functions*

The Mellin transform of a function $f(x)$, with the notation $f^*(s)$, is defined by the integral:

$$f^*(s) = \int_0^{+\infty} f(x) x^{s-1}\mathrm{d}x.$$

We shall not detail the calculus of the following transforms, but rather give Table (4.2) below of Mellin transforms, mostly determined by Reid (1995).

Table 4.2 Mellin transform of Airy functions [Reid (1995)]. We have defined $\beta = \frac{1}{6}(1-2s)$, $\gamma = \frac{1}{6}(1+s)$.

$f(x)$	$f^*(s) = \int_0^{+\infty} f(x)x^{s-1}\mathrm{d}x$
$Ai(x)$	$3^{-(s+2)/3}\frac{\Gamma(s)}{\Gamma((s+2)/3)}$, $\Re(s) > 0$
$Ai^2(x)$	$\pi^{-1/2}12^{-(2s+5)/6}\frac{2\Gamma(s)}{\Gamma((2s+5)/6)}$, $\Re(s) > 0$
$Ai(x)Bi(x)$	$2\pi^{-3/2}12^{\beta-1}\cos(\beta\pi)\Gamma(s)\Gamma(\beta)$, $0 < \Re(s) < \frac{1}{2}$
$Ai^2(-x) + Bi^2(-x)$	$4\pi^{-3/2}12^{\beta-1}\Gamma(s)\Gamma(\beta)$, $0 < \Re(s) < \frac{1}{2}$
$Ai(x)Ai(-x)$	$\pi^{-3/2}12^{\gamma-1}\sin(\gamma\pi)\Gamma(s/2)\Gamma(\gamma)$, $\Re(s) > 0$
$Ai(x)Bi(-x)$	$\pi^{-3/2}12^{\gamma-1}\cos(\gamma\pi)\Gamma(s/2)\Gamma(\gamma)$, $\Re(s) > 0$
$Ai(xe^{i\pi/6})Ai(xe^{-i\pi/6})$	$\frac{1}{2}\pi^{-3/2}12^{\gamma-1}\Gamma(s/2)\Gamma(\gamma)$, $\Re(s) > 0$

Table 4.3 Fourier transform of Airy functions

$f(x)$	$\hat{f}(\omega) = \int_{-\infty}^{+\infty} f(x)e^{-i\omega x}\mathrm{d}x$
$Ai(x)$	$\exp\left(i\omega^3/3\right)$
$Ai_1(x)$	$\left[\pi\delta(\omega) + i\wp\frac{1}{\omega}\right]\exp\left(i\omega^3/3\right)$
$Gi(x)$	$\exp\left[i\left(\omega^3/3 - \frac{\pi}{2}\mathrm{sgn}\omega\right)\right]$
$Ai(x^2)$	$2^{2/3}\pi Ai\left(-2^{-2/3}\omega\right)Ai\left(2^{-2/3}\omega\right)$
$Ai_1(x^2)$	$\frac{2\pi}{\omega}\left[Ai'\left(-2^{-2/3}\omega\right)Ai\left(2^{-2/3}\omega\right) - Ai\left(-2^{-2/3}\omega\right)Ai'\left(2^{-2/3}\omega\right)\right]$
$Ai^2(x)$	$\frac{1}{2\sqrt{\pi\|\omega\|}}\exp\left[i\left(\omega^3/12 + \frac{\pi}{4}\mathrm{sgn}\,\omega\right)\right]$
$Ai(x)Bi(x)$	$\frac{1}{2\sqrt{\pi\|\omega\|}}\exp\left[i\left(\omega^3/12 - \frac{\pi}{4}\mathrm{sgn}\,\omega\right)\right]$
$Ai(x)Ai(-x)$	$2^{-1/3}Ai\left(2^{-4/3}\omega^2\right)$
$Ai'(x)Ai'(-x)$	$2^{-1/3}\omega^2 Ai\left(2^{-4/3}\omega^2\right)$

4.3.3 *Fourier transform of Airy functions*

Table (4.3) gives some Fourier transforms of Airy functions.

4.3.4 *Hankel transform and the Airy kernel*

We define the Hankel transform of a function f by the formula (see for instance Davies (2002))

$$\check{f}(\xi) = \mathcal{H}_n\{f\}(\xi) = \int_0^\infty f(x) J_n(x\xi) x \, \mathrm{d}x, \tag{4.111}$$

and the corresponding inverse transform by

$$\mathcal{H}_n^{-1}\{\check{f}\}(x) = \int_0^\infty \check{f}(\xi) J_n(x\xi) \xi \, \mathrm{d}\xi, \tag{4.112}$$

where J_n is a Bessel function of order n.

In a paper concerning integral representations of the product of Airy functions of the form $A_2(a,b;x) = Ai(x-a)Ai(x-b), \;\; a,b \in \mathbb{R}$, Varlamov (2008-c) obtained the following result

$$A_2(a,b;x) = -2\frac{\mathrm{d}}{\mathrm{d}x} \int_0^\infty Ai^2(x - Y + \eta^2) J_0(2Z\eta) \eta \, \mathrm{d}\eta, \tag{4.113}$$

where $Y = (a+b)/2$ and $Z = (b-a)/2$. Since the Airy kernel is given by the primitive

$$\int_x^\infty A_2(a,b;t) \, \mathrm{d}t = \frac{1}{b-a}[Ai(x-a)Ai'(x-b) - Ai'(x-a)Ai(x-b)],$$

the Airy kernel may be expressed by the following Hankel transform

$$\int_x^\infty A_2(a,b;t) \, \mathrm{d}t = 2 \int_0^\infty Ai^2(x - Y + \eta^2) J_0(2Z\eta) \eta \, \mathrm{d}\eta. \tag{4.114}$$

The reader will find many other results concerning the Hankel transform in the paper by Varlamov (2008-c), in particular Hilbert transforms of Eqs. (4.113) and (4.114), where we define this transform by

$$H_x\{f(x)\} = \frac{1}{\pi} \wp \int_{-\infty}^{+\infty} \frac{f(x')}{x' - x} \mathrm{d}x'.$$

For instance, as $H_x\{Ai^2(x)\} = -Ai(x)Bi(x)$ (see Eq. (4.6)), we have

$$H_x\{A_2(a,b;x)\} = 2\frac{\mathrm{d}}{\mathrm{d}x} \int_0^\infty Ai(x - Y + \eta^2) Bi(x - Y + \eta^2) J_0(2Z\eta) \eta \, \mathrm{d}\eta. \tag{4.115}$$

Other results concerning the fractional derivatives of such transforms may also be found in the papers by Varlamov (2008-c) and Temme & Varlamov (2009).

4.4 Expansion into Fourier–Airy series

We recall that the zeros of the Airy function Ai, $\{a_n, n = 1, 2, \ldots\}$ are placed on the real negative axis of the complex plane (cf. §2.2.1). Let us consider the integral

$$I_{nn'} = \int_0^\infty Ai\,(x + a_n)\,Ai\,(x + a_{n'})\,\mathrm{d}x. \tag{4.116}$$

In the case where $n \neq n'$, formula (3.54) allows us to obtain

$$I_{nn'} = \frac{Ai\,(a_n)\,Ai'\,(a_{n'}) - Ai'\,(a_n)\,Ai\,(a_{n'})}{a_n - a_{n'}} = 0, \tag{4.117}$$

whereas if $n = n'$, formula (3.51) gives

$$I_{nn} = \int_0^\infty Ai^2\,(x + a_n)\,\mathrm{d}x = Ai'^2\,(a_n)\,. \tag{4.118}$$

In both cases, we have used the property of the Airy function to decrease exponentially toward infinity (see the asymptotic expansions (2.45) and (2.46)). Therefore the functions $\{Ai\,(x + a_n)\,/Ai'\,(a_n)\,,\ n \in \mathbb{N}\}$ form an orthonormal basis on the interval $[0, \infty)$ [Titchmarsh (1962)]. Then for any piecewise continuous integrable function $f(x)$, we can write the series

$$f(x) = \sum_{n=1}^{\infty} c_n \frac{Ai\,(x + a_n)}{Ai'\,(a_n)}. \tag{4.119}$$

The coefficients c_n of this series are determined by the relation

$$\begin{aligned} c_n &= \frac{1}{Ai'\,(a_n)} \int_0^\infty f(x) Ai\,(x + a_n)\,\mathrm{d}x \qquad (4.120) \\ &= \sum_{n=1}^{\infty} \frac{c_{n'}}{Ai'^2\,(a_n)} \int_0^\infty Ai\,(x + a_n)\,Ai\,(x + a_{n'})\,\mathrm{d}x. \end{aligned}$$

As an example, consider the function $f(x) = xAi(x)$. The above integral Eq. (4.120) gives (see formula (3.55)), for the coefficient c_n,

$$c_n = \frac{2}{a_n^3} \left[Ai(0) - a_n Ai'(0)\right].$$

Figure (4.3) illustrates the reconstitution of $f(x)$ for partial sums with 2, 4, 16 and 64 terms of the series.

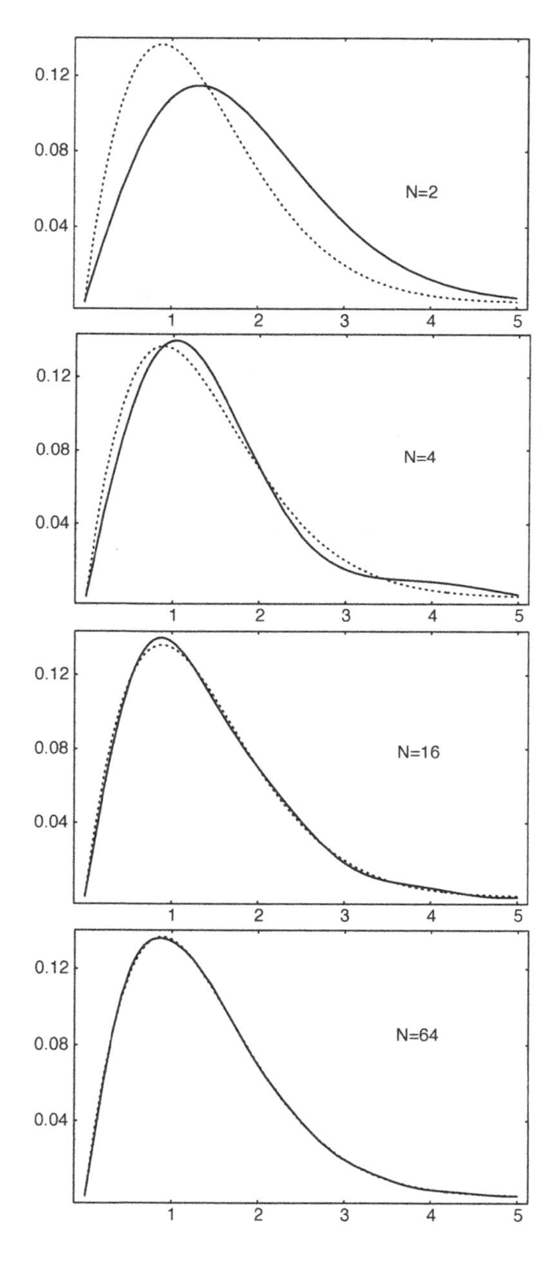

Fig. 4.3 Reconstitution of the function: $xAi(x)$ (dotted lines) by a Fourier–Airy series (solid lines) bound by 2, 4, 16 and 64 terms.

Exercises

(1) Calculate the Airy transform of $e^{i\lambda x} Ai(x)$. *Hint*: See formula (3.126).

(2) Calculate the Airy transform of the Airy wavelet $Ai'(x)Ai'(-x)$ in the case $\alpha = 1$. Is it still a wavelet? *Hint*: Use the integral (3.150), and then the property of the derivative of Airy transform (see §4.2.1).

(3) Show that the heat equation, $\partial_t u = \partial_{xx} u$, is invariant under an Airy transform. Solve the transformed equation with the initial condition $\tilde{u}_\alpha(y, 0) = \delta(y)$ i.e. the fundamental solution. Then take the inverse transform of this solution. What is the initial condition in the inverse transform $u(x, 0)$? *Hint*: See §7.4.

(4) Find the coefficients, in a Fourier–Airy expansion, of the function $f(x) = x^2 Ai(x)$. Plot the reconstitution of $f(x)$ for partial sums with 2, 4, 16 and 64 terms of the series.

Chapter 5

The Uniform Approximation

5.1 Oscillating integrals

5.1.1 *The method of stationary phase*

In this section, we will present the stationary phase method, which was studied during the early 20th century by Stokes, Kelvin and Brillouin. Erdélyi (1956) and later Copson (1967) detailed this method. A more recent review can be found in the paper by Knoll & Schaeffer (1977).

Consider an integral of the form

$$I = \int_a^b g(z) \mathrm{e}^{\mathrm{i}\lambda f(z)} \mathrm{d}z,$$

where λ is an arbitrarily large parameter.

If the function in this integral is analytic in a given domain of the complex plane, the integration contour can be deformed. The aim is to obtain an approximation of I in the limit $\lambda \to \infty$. It is then advantageous to keep the contour in regions where the integrand is as small as possible. If a topography of the complex plane with an altitude $\left|\mathrm{e}^{\mathrm{i}\lambda f(z)}\right|$ is introduced, it is then easier to speak, by analogy, of valley and top. Thus the most favourable contours will be those that remain as deep as possible in the valleys, except for the transitions from one valley to another, i.e. at points like z_i, such that: $f'(z_i) = 0$. This is the reason why this method is also called the steepest descent method. The points z_i are called the stationary points; they are the points where the integrand is maximum, giving the greatest contribution to the value of the integral I. Let us consider a function f with only one stationary point z_0 such that, $f'(z_0) = 0$. We can expand $g(z)$ to the zero order (other terms are neglected) and $f(z)$ to the second order

$$\begin{cases} g(z) \approx g(z_0) \\ f(z) \approx f(z_0) + \frac{(z-z_0)^2}{2} f''(z_0). \end{cases}$$

The integral I may be written

$$I \approx \int_a^b g(z_0) \mathrm{e}^{\frac{\mathrm{i}}{\hbar} f(z_0)} \mathrm{e}^{\mathrm{i}\lambda \frac{(z-z_0)^2}{2} f''(z_0)} \mathrm{d}z.$$

The integration can now be extended from $-\infty$ to ∞, because we only take into account the neighbourhood of the stationary point

$$I \approx g(z_0) \mathrm{e}^{\mathrm{i}\lambda f(z_0)} \int_{-\infty}^{+\infty} \mathrm{e}^{\mathrm{i}\lambda \frac{(z-z_0)^2}{2} f''(z_0)} \mathrm{d}z.$$

Applying the change of variable

$$u = (z - z_0) \left(\frac{\lambda f''(z_0)}{2\mathrm{i}} \right)^{1/2},$$

we obtain

$$I \approx \left(\frac{2\mathrm{i}}{\lambda f''(z_0)} \right)^{1/2} g(z_0) \mathrm{e}^{\mathrm{i}\lambda f(z_0)} \int_{-\infty}^{+\infty} \mathrm{e}^{-u^2} \mathrm{d}u$$

with $\int_{-\infty}^{+\infty} \mathrm{e}^{-u^2} \mathrm{d}u = \sqrt{\pi}$. This finally yields the expression

$$I \approx \left(\frac{2\pi\mathrm{i}}{\lambda f''(z_0)} \right)^{1/2} g(z_0) \mathrm{e}^{\mathrm{i}\lambda f(z_0)},$$

which can be generalised in the case where $f(z)$ admits several stationary points z_i:

$$I \underset{\lambda \to \infty}{\to} \sum_{[z_i]} g(z_i) \left[\frac{2\pi\mathrm{i}}{\lambda f''(z_i)} \right]^{1/2} \mathrm{e}^{\mathrm{i}\lambda f(z_i)}.$$

The latter expression can also be written

$$I \underset{\lambda \to \infty}{\to} \sum_{[z_i]} g(z_i) \sqrt{\frac{2\pi}{|f''(z_i)|}} \, \mathrm{e}^{\mathrm{i}\lambda \left[f(z_i) - \frac{\pi}{4} \mathrm{sgn}\left(f''(z_i) \right) \right]}.$$

5.1.2 *The uniform approximation of oscillating integrals*

The limit when $\lambda \to \infty$ of an integral of the kind [Knoll & Schaeffer (1977)]

$$I = \int g(z) e^{i\lambda f(z)} dz \tag{5.1}$$

is given in the frame of the stationary phase approximation as

$$I \underset{\lambda \to \infty}{\to} \sum_{[i]} G_i e^{i\lambda f(z_i)} = \sum_{[f'(z_i)=0]} g(z_i) \left[\frac{2\pi i}{\lambda f''(z_i)} \right]^{1/2} e^{i\lambda f(z_i)}, \tag{5.2}$$

where the stationary points z_i are defined by

$$f'(z_i) = 0.$$

This approximation is no longer valid, however, if two stationary points z_1 and z_2 coalesce, i.e. if $|f(z_1) - f(z_2)|$ is of the order $1/\lambda$. However, the uniform approximation gives uniformly valid solutions, even in the neighbourhood of coalescing stationary points. We make the change of variable

$$\tilde{f}(y) = f(z(y)),$$

such that the integrand is transformed into a simpler form allowing an analytic evaluation of the integral. The stationary points z_i become

$$\left. \frac{d\tilde{f}(y)}{dy} \right|_{y_i} = 0 \quad \Leftrightarrow \quad y_i = y(z_i),$$

and the amplitude

$$\tilde{g}(y) = g(y(z)) \frac{dz(y)}{dy}.$$

The integral Eq. (5.1) is now written in terms of the new variables

$$I = \int \tilde{g}(y) e^{i\lambda \tilde{f}(y)} dy,$$

which is tantamount to the integral (5.1), for it does not employ any approximation. The change of variables is chosen in such a manner that

$$\tilde{g}(y_i) = g(z_i) \frac{dz}{dy} = g(z_i) \left[\frac{d^2\tilde{f}}{dy^2} \Big/ \frac{d^2 f}{dz^2} \right]^{1/2}.$$

So far from the stationary points, we recover the approximation (5.2) given by the stationary phase method.

We will now study a particular case of the uniform treatment of the integral (5.1): the uniform approximation by Airy function.

5.1.3 *The Airy uniform approximation*

Numerous variations on this method can be found in the scientific literature, for instance Child (1974) or Knoll & Schaeffer (1977). For $\tilde{f}$, we choose the cubic form

$$\tilde{f}(y) = f\left(z(y)\right) = \lambda\left(\eta - \xi^2 y - \frac{y^3}{3}\right), \tag{5.3}$$

with the stationary points $y_{1,2} = \pm\xi$. Taking Eq. (5.3) in z_1 and z_2, we obtain

$$\begin{cases} \eta = \lambda f\left(z_1\right) + f\left(\frac{z_2}{2}\right) \\ \xi = \lambda\left[\frac{3}{4}\left(f\left(z_1\right) - f\left(z_2\right)\right)\right]^{1/3}. \end{cases}$$

Now if the a linear form is taken for the amplitude, we have

$$\tilde{g}(y) = r_1 + \mathrm{i}r_2,$$

so that we obtain

$$I \to I_{Ai} = \left[r_1 Ai\left(-\xi^2\right) + r_2 Ai'\left(-\xi^2\right)\right]\mathrm{e}^{\mathrm{i}\eta}, \tag{5.4}$$

where r_1 and r_2 are defined in such a manner that the integrals (5.1) and (5.4) have the same asymptotic behaviour when $\lambda \to \infty$. That is to say

$$\begin{cases} r_1 + \mathrm{i}\xi r_2 = 2\sqrt{\pi\xi}G_1\mathrm{e}^{-\mathrm{i}\pi/4} \\ \\ r_1 - \mathrm{i}\xi r_2 = 2\sqrt{\pi\xi}G_2\mathrm{e}^{\mathrm{i}\pi/4}, \end{cases}$$

with amplitudes G_1 and G_2 defined by Eq. (5.2)

$$G_i = g(z_i)\left[\frac{2\pi\mathrm{i}}{f''(z_i)}\right]^{1/2}.$$

5.2 Differential equations of the second order

5.2.1 *The JWKB method*

Liouville (1837) and Green (1837) were looking for a solution to the second order differential equation [Morse & Feshbach (1953); Berry & Mount (1972); Nayfeh (1973); Olver (1974)]:

$$\frac{\mathrm{d}^2y}{\mathrm{d}x^2} + \left[\lambda^2 q_1(x) + q_2(x)\right]y = 0, \tag{5.5}$$

for large values of λ, where $q_1(x)$ is a positive, twice differentiable function on a given interval. The function $q_2(x)$ is continuous in the same interval. Making the following transforms on the dependent and the independent variables

$$\begin{cases} z = \phi(x) \\ v = \psi(x)y(x), \end{cases}$$

the differential equation (5.5) becomes

$$\frac{\mathrm{d}^2 v}{\mathrm{d}z^2} + \frac{1}{\phi'^2}\left(\phi'' - \frac{2\phi'\psi'}{\psi}\right)\frac{\mathrm{d}v}{\mathrm{d}z} + \frac{1}{\phi'^2}\left[\lambda^2 q_1(x) + q_2(x) - \psi\left(\frac{\psi'}{\psi^2}\right)'\right]v = 0. \tag{5.6}$$

In order to have a self-similar equation in the transformation, the second term must be cancelled, namely

$$\phi'' - \frac{2\phi'\psi'}{\psi} = 0,$$

or

$$\psi = \sqrt{\phi'}. \tag{5.7}$$

Eq. (5.6) then becomes

$$\frac{\mathrm{d}^2 v}{\mathrm{d}z^2} + \left[\lambda^2 \frac{q_1}{\phi'^2} + \frac{q_2}{\phi'^2} - \frac{1}{2\phi'^2}\{\phi, x\}\right]v = 0, \tag{5.8}$$

where $\{\phi, x\}$ is the Schwarzian derivative of ϕ:

$$\{\phi, x\} = \frac{\phi'''}{\phi'} - \frac{3}{2}\left(\frac{\phi''}{\phi'}\right)^2. \tag{5.9}$$

Note that Eq. (5.8) is strictly equivalent to Eq. (5.5), because no approximation was necessary to obtain the result. If we now assume that λ is large enough, we can neglect the last two terms inside the square brackets of Eq. (5.8). Consequently Eq. (5.8) becomes, in a first approximation,

$$\frac{\mathrm{d}^2 v}{\mathrm{d}z^2} + \lambda^2 \frac{q_1}{\phi'^2} v = 0. \tag{5.10}$$

Then we choose ϕ such that

$$\frac{q_1}{\phi'^2} = 1, \tag{5.11}$$

that is to say

$$\phi = \int \sqrt{q_1(t)}\mathrm{d}t. \tag{5.12}$$

Eq. (5.10) now reads

$$\frac{\mathrm{d}^2 v}{\mathrm{d}z^2} + \lambda^2 v = 0. \tag{5.13}$$

The solutions to the differential equation (5.13) are

$$v = a\cos(\lambda z) + b\sin(\lambda z),$$

so, in making the inverse transform, we find[1]

$$y = \frac{1}{[q_1(x)]^{1/4}}\left\{a\cos\left(\lambda\int\sqrt{q_1(x)}\mathrm{d}x\right) + b\sin\left(\lambda\int\sqrt{q_1(x)}\mathrm{d}x\right)\right\}. \tag{5.14}$$

But if the function $q_1(x)$ is negative, the solution is

$$y = \frac{1}{[-q_1(x)]^{1/4}}\left\{a\mathrm{e}^{\lambda\int\sqrt{-q_1(x)}\mathrm{d}x} + b\mathrm{e}^{-\lambda\int\sqrt{-q_1(x)}\mathrm{d}x}\right\}. \tag{5.15}$$

We should mention that this approximation is no longer valid in the neighbourhood of the zeros of $q_1(x)$, i.e. in the neighbourhood of the, so called, turning points. The transformations (5.7) and (5.12) are called the Green–Liouville transformations. The solutions (5.14) and (5.15) are called JWKB approximated solutions, after Jeffreys (1923), Wentzel (1926), Kramers (1926) and Brillouin (1926).

5.2.2 *The Langer generalisation*

The factor $q_1^{-1/4}(x)$, in the solutions (5.14) and (5.15) of Eq. (5.5), yields the divergence of these solutions at the turning points. In order to have a uniformly valid solution in the whole neighbourhood of a turning point x_t, we put [Langer (1931)]

$$\frac{q_1}{\phi'^2} = \phi, \tag{5.16}$$

in other words

$$\begin{cases} \frac{2}{3}\phi^{3/2} = \int\limits_{x_t}^{x} \sqrt{q_1(t)}\,\mathrm{d}t & \text{if } x > x_t \\ \frac{2}{3}(-\phi)^{3/2} = \int\limits_{x}^{x_t} \sqrt{-q_1(t)}\,\mathrm{d}t & \text{if } x < x_t. \end{cases} \tag{5.17}$$

[1] A convenient phase factor is introduced usually leading to the solution: $y = N[q_1(x)]^{-1/4}\cos\left(\lambda\int\sqrt{q_1(x)}\mathrm{d}x - \frac{\pi}{4}\right)$, where N is a normalisation constant.

The differential equation (5.10) then becomes

$$\frac{\mathrm{d}^2 v}{\mathrm{d}z^2} + \lambda^2 z v = 0, \tag{5.18}$$

whose solutions are

$$v = aAi\left(-\lambda^{2/3} z\right) + bBi\left(-\lambda^{2/3} z\right),$$

where Ai and Bi are the homogeneous Airy functions. The uniform solution to Eq. (5.5), valid even in the neighbourhood of the turning point, is

$$y = \frac{1}{\sqrt{\phi'(x)}}\left\{aAi\left[-\lambda^{2/3}\phi(x)\right] + bBi\left[-\lambda^{2/3}\phi(x)\right]\right\}. \tag{5.19}$$

We can see that, in the asymptotic limit, the expression (5.19) goes to the JWKB approximation given by formulae (5.14) and (5.15), thanks to (2.45) and (2.47).

Olver (1954) generalises the Langer transformation, setting

$$\begin{cases} \zeta = \zeta(z) = \int\limits^{x} \sqrt{q_1(t)}\,\mathrm{d}t \\ \chi = \frac{q_1(x)}{\zeta'^2}, \quad y = \chi^{-1/4} v, \end{cases}$$

where the independent variable z is an arbitrary function of x. Equation (5.5) then reads

$$\frac{\mathrm{d}^2 v}{\mathrm{d}z^2} + \lambda^2 \zeta'^2 v = 0.$$

The solutions of this equation are asymptotically equivalent to the solutions (5.14) and (5.15). Consequently, $\zeta(z)$ has to be chosen in such a manner that ζ'^2 has the same number of zeros as $q_1(x)$. For example, if $q_1(x)$ has two turning points, we have [Pike (1964)]:

$$\zeta'^2 = 4a^2\left(1 - z^2\right),$$

such that $z = -1$ corresponds to $x = x_1$, and $z = +1$ to $x = x_2$. We then obtain

$$\zeta = 2a \int\limits_{-1}^{z} \sqrt{1 - t^2}\mathrm{d}t = \int\limits_{x_1}^{x} \sqrt{q_1(t)}\mathrm{d}t,$$

that is to say

$$a = \frac{1}{\pi} \int\limits_{x_1}^{x_2} \sqrt{q_1(t)}\mathrm{d}t.$$

Equation (5.8) now reads

$$\frac{\mathrm{d}^2 v}{\mathrm{d}z^2} + 4a^2\lambda^2\left(1 - z^2\right)v = 0,$$

with the solution

$$v = D_\mu\left(2\sqrt{a\lambda z}\right),\ \mu + \frac{1}{2} = a\lambda,$$

where D_μ is the parabolic cylinder function of order μ [Abramowitz & Stegun (1965)].

5.3 Inhomogeneous differential equations

Consider the following inhomogeneous differential equation

$$\frac{\mathrm{d}^2 y}{\mathrm{d}x^2} + \left[\lambda^2 q_1(x) + q_2(x)\right] y = \lambda^2 G(x), \tag{5.20}$$

where $q_1(x)$ has a simple zero at $x = x_t$. When the parameter λ goes to infinity, we obtain a particular approximate solution as

$$y = \frac{G(x)}{q_1(x)}.$$

This solution is singular for $x = x_t$, except if $G(x)$ also has a zero for $x = x_t$. In order to determine a particular solution in the case where $G(x_t) \neq 0$, we use the transformation

$$\begin{cases} z = \phi(x) \\ \frac{2}{3} z^{3/2} = \int\limits_{x_t}^{x} \sqrt{q_1(t)}\mathrm{d}t\, y = \frac{v}{\sqrt{\phi'}}. \end{cases}$$

Equation (5.20) now reads

$$\frac{\mathrm{d}^2 v}{\mathrm{d}z^2} + \left[\lambda^2 z - \delta\right] v = \lambda^2 g(z), \tag{5.21}$$

where δ is defined by the relation

$$\delta = \frac{1}{2\phi'^2}\{\phi, x\} - \frac{q_2}{\phi'^2},$$

with $g(z) = \{\phi'[x(z)]\}^{-3/2} G[x(z)]$, $\{\phi, x\}$ being the Schwarzian derivative of ϕ relative to x. If δ is negligible compared to $\lambda^2 z$, Eq. (5.21) becomes:

$$\frac{\mathrm{d}^2 v}{\mathrm{d}z^2} + \lambda^2 z v = \lambda^2 g(z). \tag{5.22}$$

Therefore, we can write $g(z)$ as the sum of two terms

$$g(z) = g(0) + [g(z) - g(0)],$$

and hence, we have to determine a particular solution for each of these terms. A particular solution for the second term is given in a first approximation by [Nayfeh (1973)]:

$$v_1 = \frac{g(z) - g(0)}{z}.$$

In order to find a particular solution corresponding to the first term, we put $\xi = \lambda^{2/3} z$. Equation (5.22) now reads

$$\frac{\mathrm{d}^2 v}{\mathrm{d}\xi^2} + \xi v = \lambda^{2/3} g(0),$$

with the particular solution

$$v_2 = \pi \lambda^{2/3} g(0) Wi(\xi),$$

where Wi is an inhomogeneous Airy function, either Hi or Gi, depending on whether the sign of $g(0)$ is positive or negative respectively. A particular solution to (5.22) is then

$$v = \pi \lambda^{2/3} g(0) Wi(\xi) + \frac{g(z) - g(0)}{z}.$$

From the inverse transform, we obtain

$$\begin{aligned} y = {} & \pi \lambda^{2/3} \frac{G(x_t)}{[\phi'(x) f(x_t)]^{1/2}} Wi\left(\lambda^{2/3} z\right) \\ & + \frac{1}{q_1(x)} \left[G(x) - \frac{G(x_t) \phi^{3/2}}{\sqrt{f(x_t)}} \right] \end{aligned} \tag{5.23}$$

with $f(x) = \frac{q(x_t)}{x - x_t}$.

Exercises

Consider the following nonlinear third order differential equation:

$$\dddot{x} + a\,\ddot{x} - x\dot{x} + x = 0.$$

(1) Apply a scaling on x and t to transform this equation into

$$\dddot{x} + \lambda^2(\ddot{x} - x\dot{x} + x) = 0.$$

(2) In the limit $\lambda \to \infty$, prove that a solution $x(t) + \mathcal{O}(1/\lambda)$ satisfies the second order differential equation, inside the parenthesis of the above equation.

(3) Find a particular solution $x_p(t)$ of the second order equation. Check that

$$I = (y-1)\exp\left(y - \frac{x^2}{2a}\right)$$

is a first integral of this equation ($y = \dot{x}$). Plot this first integral in the phase space for different values of $I =$ constant.

(4) Using the transformation $x \to t + x_0 + u$ where x_0 is a constant and u is a perturbed solution searched near the particular solution, find the third order linear equation satisfied by u from the initial third order equation.

(5) Solve this equation in terms of Airy functions.

(6) What is the condition on the integration constants, for which the solution is a true perturbed solution?

Hint: The solution to this problem is given in [Letellier & Vallée (2003)].

Chapter 6

Generalisation of Airy Functions

6.1 Generalisation of the Airy integral

6.1.1 *The generalisation of Watson*

Airy functions are solutions to the Airy differential equation $y'' - xy = 0$. A generalisation of these functions may be made by considering the solutions to second order differential equations of the kind: $y'' + \mathrm{C}x^k y = 0$.

Since the Airy function is defined by the integral, (2.20)

$$Ai(x) = \frac{1}{\pi} \int_0^\infty \cos\left(\frac{z^3}{3} + xz\right) \mathrm{d}z,$$

we can generalise this integral as follows. Following Watson (1966), we put

$$T_n(t,\alpha) = t^n \, F\left(-\frac{n}{2}, \frac{1-n}{2}; 1-n; -\frac{4\alpha}{t^2}\right), \tag{6.1}$$

where $n \in \mathbb{N}$, $n \geq 2$, and F is the hypergeometric function defined by [Abramowitz & Stegun (1965)]

$$F(a,b;c;z) = \sum_{n=0}^{\infty} \frac{(a)_n (b)_n}{(c)_n} \frac{z^n}{n!}.$$

Thus we obtain

$$\begin{aligned}
T_2(t,\alpha) &= t^2 + 2\alpha \\
T_3(t,\alpha) &= t^3 + 3\alpha t \\
T_4(t,\alpha) &= t^4 + 4\alpha t^2 + 2\alpha^2 \\
T_5(t,\alpha) &= t^5 + 5\alpha t^3 + 5\alpha^2 t \\
\vdots \quad & \quad \vdots
\end{aligned} \tag{6.2}$$

So the generalisation of the Airy integral is given by:

$$Ci_n(\alpha) = \int_0^\infty \cos\left[T_n\left(t, \alpha\right)\right] \mathrm{d}t \tag{6.3}$$

$$Si_n(\alpha) = \int_0^\infty \sin\left[T_n\left(t, \alpha\right)\right] \mathrm{d}t \tag{6.4}$$

$$Ei_n(\alpha) = \int_0^\infty \exp\left[-T_n\left(t, \alpha\right)\right] \mathrm{d}t. \tag{6.5}$$

As particular cases, we have $Ai(x) = \frac{3^{1/3}}{\pi} Ci_3\left(\frac{x}{3^{2/3}}\right)$ (formula (2.20)) and $Gi(x) = \frac{3^{1/3}}{\pi} Si_3\left(\frac{x}{3^{2/3}}\right)$ (formula (2.127)).[1]

As in the case of Airy functions (cf. §2.2.4), we can express the integrals (6.3), (6.4) and (6.5) thanks to the Bessel functions I, J and K, according to the parity of n.

We shall not detail the calculations (see for instance Watson (1966)), but if n is even, $Ci_n(\alpha)$ and $Si_n(\alpha)$ are solutions to the differential equation

$$\frac{\mathrm{d}^2 Wi}{\mathrm{d}\alpha^2} + n^2 \alpha^{n-2} Wi = 0, \tag{6.6}$$

with $Wi = Ci_n(\alpha)$, $Si_n(\alpha)$. The function $Ei_n(\alpha)$ is a solution to the equation

$$\frac{\mathrm{d}^2 Ei_n}{\mathrm{d}\alpha^2} - n^2 \alpha^{n-2} Ei_n = 0. \tag{6.7}$$

These three functions may be expressed in the following form, with $\alpha > 0$,

$$Ci_n(\alpha) = \frac{\pi \alpha^{1/2}}{2n \sin\left(\frac{\pi}{2n}\right)} \left[J_{-1/n}\left(2\alpha^{n/2}\right) - J_{1/n}\left(2\alpha^{n/2}\right)\right] \tag{6.8}$$

$$Si_n(\alpha) = \frac{\pi \alpha^{1/2}}{2n \sin\left(\frac{\pi}{2n}\right)} \left[J_{-1/n}\left(2\alpha^{n/2}\right) + J_{1/n}\left(2\alpha^{n/2}\right)\right] \tag{6.9}$$

$$Ei_n(\alpha) = \frac{2\alpha^{1/2}}{n} K_{1/n}\left(2\alpha^{n/2}\right), \tag{6.10}$$

and

$$Ci_n(-\alpha) = \frac{\pi \alpha^{1/2}}{2n \sin\left(\frac{\pi}{2n}\right)} \left[J_{-1/n}\left(2\alpha^{n/2}\right) + J_{1/n}\left(2\alpha^{n/2}\right)\right] \tag{6.11}$$

$$Si_n(-\alpha) = \frac{\pi \alpha^{1/2}}{2n \sin\left(\frac{\pi}{2n}\right)} \left[J_{-1/n}\left(2\alpha^{n/2}\right) - J_{1/n}\left(2\alpha^{n/2}\right)\right] \tag{6.12}$$

$$Ei_n(-\alpha) = \frac{\pi \alpha^{1/2}}{n \sin\left(\frac{\pi}{n}\right)} \left[I_{-1/n}\left(2\alpha^{n/2}\right) + I_{1/n}\left(2\alpha^{n/2}\right)\right]. \tag{6.13}$$

[1]Note that the functions Ci, Si and Ei (so defined) have nothing to do with the functions: cosine integral, sine integral and exponential integral.

Similarly, if n is odd, we obtain for $\alpha > 0$,

$$Ci_n(\alpha) = \frac{\pi\alpha^{1/2}}{2n\sin\left(\frac{\pi}{2n}\right)}\left[I_{-1/n}\left(2\alpha^{n/2}\right) - I_{1/n}\left(2\alpha^{n/2}\right)\right] \tag{6.14}$$

$$= \frac{2\alpha^{1/2}\cos\left(\frac{\pi}{2n}\right)}{n} K_{1/n}\left(2\alpha^{n/2}\right) \tag{6.15}$$

$$Ci_n(-\alpha) = \frac{\pi\alpha^{1/2}}{2n\sin\left(\frac{\pi}{2n}\right)}\left[J_{-1/n}\left(2\alpha^{n/2}\right) + J_{1/n}\left(2\alpha^{n/2}\right)\right]. \tag{6.16}$$

The function $Ci_n(\alpha)$ (n odd) then verifies the differential equation

$$\frac{\mathrm{d}^2 Ci_n}{\mathrm{d}\alpha^2} - n^2\alpha^{n-2} Ci_n = 0. \tag{6.17}$$

In particular for $n = 3$, we find the Airy equation (2.1).

The expressions for Si and Ei are more complicated. $Si_n(-\alpha)$ can be written in the form ($\alpha > 0$)

$$Si_n(\alpha) = -\frac{\pi\alpha^{\frac{n+1}{2}}}{n\cos\left(\frac{\pi}{2n}\right)}\sum_{m=0}^{\infty}\frac{\alpha^{mn}}{\Gamma\left(m+\frac{3}{2}-\frac{1}{2n}\right)\Gamma\left(m+\frac{3}{2}+\frac{1}{2n}\right)} \tag{6.18}$$

$$+\frac{\pi\alpha^{1/2}}{2n\cos\left(\frac{\pi}{2n}\right)}\left[I_{-1/n}\left(2\alpha^{n/2}\right) + I_{1/n}\left(2\alpha^{n/2}\right)\right]$$

$$Si_n(-\alpha) = \frac{(-1)^{\frac{n-1}{2}}\pi\alpha^{\frac{n+1}{2}}}{n\cos\left(\frac{\pi}{2n}\right)}\sum_{m=0}^{\infty}\frac{(-1)^m\alpha^{mn}}{\Gamma\left(m+\frac{3}{2}-\frac{1}{2n}\right)\Gamma\left(m+\frac{3}{2}+\frac{1}{2n}\right)}$$

$$+\frac{\pi\alpha^{1/2}}{2n\cos\left(\frac{\pi}{2n}\right)}\left[J_{-1/n}\left(2\alpha^{n/2}\right) - J_{1/n}\left(2\alpha^{n/2}\right)\right]. \tag{6.19}$$

$Si_n(\alpha)$ (n odd) satisfies the differential equation

$$\frac{\mathrm{d}^2 Si_n}{\mathrm{d}\alpha^2} - n^2\alpha^{n-2} Si_n = -n\alpha^{\frac{n-3}{2}}. \tag{6.20}$$

In particular for $n = 3$, we again find the Airy inhomogeneous differential equation (2.126).

The function $Ei_n(\alpha)$ (n odd) may be written as the series

$$Ei_n(\alpha) = \frac{\pi\alpha^{\frac{n+1}{2}}}{n\cos\left(\frac{\pi}{2n}\right)}\sum_{m=0}^{\infty}\frac{(-1)^m\alpha^{mn}}{\Gamma\left(m+\frac{3}{2}-\frac{1}{2n}\right)\Gamma\left(m+\frac{3}{2}+\frac{1}{2n}\right)} \tag{6.21}$$

$$+\frac{\pi}{n\sin\left(\frac{\pi}{n}\right)}\left\{\sum_{m=0}^{\infty}\frac{(-1)^m\alpha^{mn}}{m!\Gamma\left(m+1-\frac{1}{n}\right)}\right.$$

$$\left.-\alpha\sum_{m=0}^{\infty}\frac{(-1)^m\alpha^{mn}}{m!\Gamma\left(m+1+\frac{1}{n}\right)}\right\}.$$

$Ei_n(\alpha)$ (n odd) satisfies the differential equation

$$\frac{\mathrm{d}^2 Ei_n}{\mathrm{d}\alpha^2} + n^2\alpha^{n-2} Ei_n = n\alpha^{\frac{n-3}{2}}. \tag{6.22}$$

6.1.2 Oscillating integrals and catastrophes

The Watson's method involves oscillating integrals that generalise the Airy integral. However, these integrals depend only on one parameter, which limits this generalisation to a certain extent. Another way to approach the problem is the use of "catastrophe theory" and the associated oscillating integrals [Thom (1975); Arnold (1981); Poston & Stewart (1978)]. In a series of articles devoted to the subject, Berry (and co-workers) deals with the problem of the relationship between caustics, diffraction phenomena and catastrophe theory [see for instance: Berry (and co-workers) (1976), (1979), (1980a), (1980b)]. Briefly summarized, catastrophe theory consists of the classification of structurally stable singularities. Structural stability means singularities that are unchanged under small perturbations.

In the following table (Table (6.1)), we give the first elementary catastrophes in terms of the dimension, D, they unfold, and as a function of the number of K parameters for their description, which is called the codimension.

Table 6.1 The five elementary catastrophes with a codimension $K \leq 3$. $\mathbf{V}$ is the parameter vector in the K-dimensional space.

K	Name	Canonical polynomial
	Cuspoids $D = 1$	$\phi(\xi; \mathbf{V})$
1	Fold	$\xi^3/3 + x\,\xi$
2	Cusp	$\xi^4/4 + x\,\xi^2/2 + y\,\xi$
3	Swallowtail	$\xi^5/5 + z\,\xi^3/3 + y\,\xi^2/2 + x\,\xi$
	Umbilics $D = 2$	$\phi(\xi, \eta; \mathbf{V})$
3	Elliptic umbilic	$\xi^3 - 3\xi\eta^2 - z(\xi^2 + \eta^2) - x\,\xi - y\,\eta$
3	Hyperbolic umbilic	$\xi^3 + \eta^3 + z\,\xi\eta - x\,\xi - y\,\eta$

We now associate oscillating integrals with the catastrophe polynomials $\phi(\xi;\mathbf{V})$,

$$C_k(\mathbf{V}) = \frac{1}{\sqrt{2\pi}} \int_{-\infty}^{+\infty} \exp[\mathrm{i}\phi_k(\xi;\mathbf{V})]\,\mathrm{d}\xi,$$

for the cuspoids, and for the umbilics

$$U_k(\mathbf{V}) = \frac{1}{2\pi} \int_{-\infty}^{+\infty}\int_{-\infty}^{+\infty} \exp[\mathrm{i}\phi_k(\xi,\eta;\mathbf{V})]\,\mathrm{d}\xi\mathrm{d}\eta.$$

So the fold catastrophe is associated with an oscillating integral, which is proportional to the Airy function:

$$C_1(x) = \frac{1}{\sqrt{2\pi}} \int_{-\infty}^{+\infty} \exp[\mathrm{i}(\xi^3/3 + x\,\xi)]\,\mathrm{d}\xi = \sqrt{2\pi}Ai(x),$$

while for the cusp catastrophe, we have

$$C_2(x,y) = \frac{1}{\sqrt{2\pi}} \int_{-\infty}^{+\infty} \exp[\mathrm{i}(\xi^4/4 + x\,\xi^2/2 + y\,\xi)]\,\mathrm{d}\xi,$$

C_2 is related to the Pearcey function [Pearcey (1946)] which pioneered the mathematical description of the cusp of a caustic (see also Connor & Farrelly (1980)). Moreover, when $y = 0$, this function is connected to the method of Watson, for the case $n = 4$.

Finally the swallowtail function is naturally given by

$$C_3(x,y,z) = \frac{1}{\sqrt{2\pi}} \int_{-\infty}^{+\infty} \exp[\mathrm{i}(\xi^5/5 + z\,\xi^3/3 + y\,\xi^2/2 + x\,\xi)]\,\mathrm{d}\xi.$$

For the umbilics, we first have the elliptic umbilic function

$$E(x,y,z) = \frac{1}{2\pi} \int_{-\infty}^{+\infty}\int_{-\infty}^{+\infty} \exp[\mathrm{i}(\xi^3 - 3\xi\eta^2 - z(\xi^2+\eta^2) - x\,\xi - y\,\eta)]\,\mathrm{d}\xi\mathrm{d}\eta,$$

and the hyperbolic umbilic function

$$H(x,y,z) = \frac{1}{2\pi} \int_{-\infty}^{+\infty}\int_{-\infty}^{+\infty} \exp[\mathrm{i}(\xi^3 + \eta^3 + z\,\xi\eta - x\,\xi - y\,\eta)]\,\mathrm{d}\xi\mathrm{d}\eta.$$

There is a direct relation between these catastrophe functions and the Airy function itself.

In fact, the Airy function satisfies the relation

$$Ai^2(x) = \frac{1}{2^{1/3}} \int_{-\infty}^{+\infty} Ai[2^{2/3}(x+u^2)]\,du,$$

(see for instance Balazs & Zipfel (1973), or Berry (1977a)) by projecting a quantum mechanical Wigner function from the phase space onto coordinate space. The method described by Berry and Wright (1980) is a generalisation of this result. For example, they obtain relations between the cusp function and the fold function:

$$|C_2(x,y)|^2 = \sqrt{\frac{2}{\pi}} \int_0^{+\infty} \Big\{ C_1[2(6u)^{-1/3}(u^3+yu+x)] \\ + C_1[2(6u)^{-1/3}(u^3+yu-x)] \Big\} (6u)^{-1/3}\,du,$$

and

$$|C_2(x,y)|^2 = 2\sqrt{\frac{2}{\pi}}\,\Re \int_0^{+\infty} (6v)^{-1/3} e^{2ixv}\, C_1[2(6v)^{-1/3}(v^3+yv)]\,dv.$$

The latter relation may be expressed in terms of the Airy function since with the change $v = t^{3/2}$, we have

$$|C_2(x,y)|^2 = 6^{2/3} \int_0^{+\infty} Ai\left[\left(\frac{4}{3}\right)^{1/3} (t^4 + yt) \right] \cos(2xt^{3/2})\,dt. \qquad (6.23)$$

We have plotted the square of the cusp function (Eq. (6.23) on Fig. (6.1) which depicts, for instance, the intensity of light in the neighbourhood of the cusp of a caustic, in the same way that the square of the Airy function depicts the intensity of light along the caustic of a rainbow. By the same method, there is a connection between the square of the modulus of the swallowtail function and the cusp function. Moreover there are similar relations for the umbilics. For the elliptic umbilic, we have the identity (Berry & Wright (1980))

$$|E(x,y,z)|^2 = \frac{2^{1/3}}{\pi} \int_{-\infty}^{+\infty} d\xi \int_{-\infty}^{+\infty} d\eta \\ E\left\{ 2^{2/3}[x + 2z\xi + 3(\eta^2 - \xi^2)],\, 2^{2/3}(y + 2z\eta^2 + 6\xi\eta),\, 0 \right\}, \qquad (6.24)$$

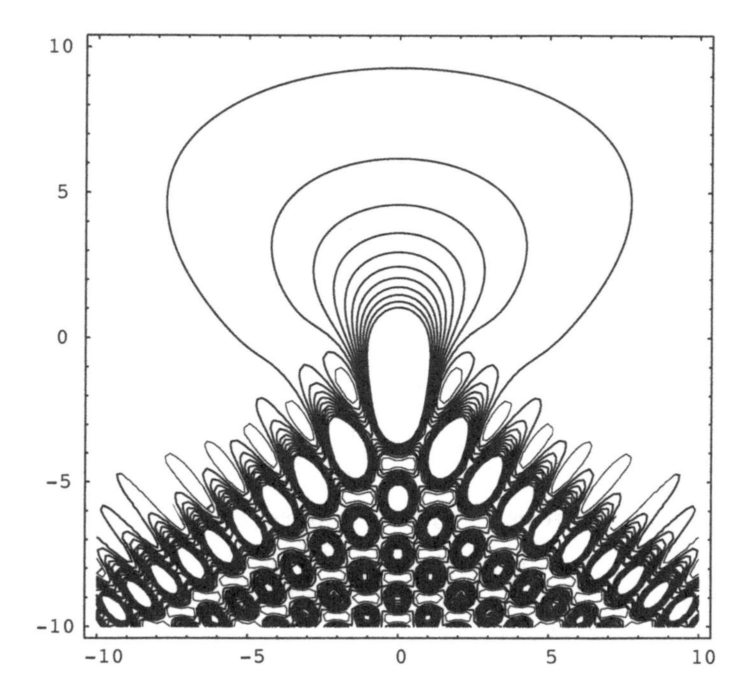

Fig. 6.1 Square of the modulus of the cusp catastrophe function.

which can be written in terms of Airy functions thanks to Eq. (3.125),

$$E(u,v,0) = \left(\frac{2}{3}\right)^{2/3} \pi \, \Re\left[Ai\left(\frac{-u-\mathrm{i}v}{12^{1/3}}\right) Bi\left(\frac{-u+\mathrm{i}v}{12^{1/3}}\right)\right],$$

and for the hyperbolic umbilic, the identity

$$|H(x,y,z)|^2 = \frac{2^{1/3}}{\pi} \int_{-\infty}^{+\infty} \mathrm{d}\xi \int_{-\infty}^{+\infty} \mathrm{d}\eta H[2^{2/3}(x-z\eta-3\xi^2), 2^{2/3}(y-z\xi-3\eta^2), 0], \tag{6.25}$$

which can be written in terms of Airy functions thanks to the relation

$$H(u,v,0) = \left(\frac{2\pi}{3^{2/3}}\right) Ai(-3^{1/3}u) Ai(-3^{1/3}v).$$

Details and other relationships may be found in the paper by Berry and Wright (1980).

All these results show that catastrophe oscillating integrals may be considered as generalisations of Airy functions.

6.2 Third order differential equations

6.2.1 *The linear third order differential equation*

This section gives some generalities about the linear third order differential equation

$$y''' + P(x)y'' + Q(x)y' + R(x)y = 0. \tag{6.26}$$

This can be written in the canonical form [Ince (1956)]

$$z''' + f(x)z' + g(x)z = 0, \tag{6.27}$$

by making following change of function

$$y(x) = z(x)\exp\left(-\frac{1}{3}\int^x P(t)\mathrm{d}t\right).$$

If L is the operator associated with the above equation

$$L[z] = z''' + f(x)z' + g(x)z, \tag{6.28}$$

then the adjoint operator $\overline{L}$ reads

$$\overline{L}[y] = -y''' - \frac{\mathrm{d}}{\mathrm{d}x}(f(x)y) + g(x)y. \tag{6.29}$$

Now we have the following theorem:

Theorem 6.1. *If z_i $(i = 1, 2, 3)$ are the three linearly independent solutions of the equation $L[z] = 0$, then the adjoint equation $\overline{L}[z] = 0$ admits, as solutions, the minors of the Wronskian of the equation $L[z] = 0$: $y_i = \epsilon_{ijk}(z_j z_k' - z_j' z_k)$, where ϵ_{ijk} is the completely antisymmetric tensor.*

The Wronskian of the equation $L[z] = 0$ is defined by the determinant

$$W(z_1, z_2, z_3) = \begin{vmatrix} z_1 & z_2 & z_3 \\ z_1' & z_2' & z_3' \\ z_1'' & z_2'' & z_3'' \end{vmatrix} = \mathrm{C}^{\mathrm{t}}.$$

The proof of this theorem is given by a simple substitution. More likely the result may be generalised to the canonical differential equation of any order. In particular, it is clearly seen in the case of the second order differential equation $z''+h(x)z = 0$ (which is self-adjoint), where the minors are written as $y_i = \delta_{ij}z_j$, $(i, j = 1, 2)$.

6.2.2 *Asymptotic solutions*

Let us consider the differential equation [Langer (1955a,b)]

$$z''' + \lambda^2 \left(f(x)z' + g(x)z\right) = 0, \tag{6.30}$$

where λ is a large parameter. We are looking for the solutions to this equation when $\lambda \to \infty$, and when $f(x)$ has a transition point x_0, i.e. when $f(x_0) = 0$. Moreover, we assume that $f'(x_0) \neq 0$ and $g(x_0) \neq 0$. We start with a change of the dependent and independent variables, as was done for the second order differential equation (cf. §5.2)

$$\begin{cases} u = u(x) \\ z = \frac{1}{u'}\phi(u). \end{cases}$$

Equation (6.30) thus becomes

$$\phi''' + \left[\lambda^2 \frac{f\left(x(u)\right)}{u'^2} + S(u)\right]\phi' + \left[\lambda^2 \frac{g\left(x(u)\right)}{u'^3} + \frac{1}{2}\frac{\mathrm{d}S(u)}{\mathrm{d}u}\right]\phi = 0, \tag{6.31}$$

where ϕ' is the derivative of ϕ with respect to the variable u and u' the derivative of u with respect to the variable x. The quantity $S(u)$ is proportional to the Schwarzian derivative, $\{u, x\}$

$$S(u) = -\frac{1}{u'^2}\{u, x\} = -\frac{1}{u'^2}\left[\frac{u'''}{u'} - \frac{3}{2}\left(\frac{u''}{u'}\right)^2\right].$$

Now we complete the choice of the change of variable, setting $u = f(x)/u'^2$, namely

$$u = \left[\frac{3}{2}\int_{x_0}^{x} f(x')^{\frac{1}{2}}\mathrm{d}x'\right]^{\frac{2}{3}}. \tag{6.32}$$

Equation (6.31) thus becomes

$$\phi''' + \left[\lambda^2 u + S(u)\right]\phi' + \left[\lambda^2 h(u) + \frac{1}{2}\frac{\mathrm{d}S(u)}{\mathrm{d}u}\right]\phi = 0, \tag{6.33}$$

with $h(u) = g\left(x(u)\right)/u'^3$. However, we have not used the asymptotic limit $\lambda \to \infty$, which allows the problem to be simplified. In particular, we may compare Eq. (6.33) to the following reference (comparison) equation, where $\mu =$ constant,

$$\Phi''' + \lambda^2 u\Phi' + \lambda^2 \mu\Phi = 0. \tag{6.34}$$

For this purpose, we carry out the asymptotic expansion $\phi(u) = \varpi(u)\Phi(u) + \mathcal{O}\left(\frac{1}{\lambda}\right)$, which is introduced into Eq. (6.34), leading (up to terms $\mathcal{O}\left(\frac{1}{\lambda}\right)$) to the identification

$$u\varpi'(u) = \left(\mu - h(u)\right)\varpi(u). \tag{6.35}$$

After integrating, we find

$$\varpi(u) = \exp\left[-\int_0^u \frac{h(v)-\mu}{v}\mathrm{d}v\right]. \tag{6.36}$$

The convergence of the integral in the relation (6.36) is ensured by setting $h(0) = \mu$, allowing the asymptotic solution to Eq. (6.30) to be written as

$$\lim_{\lambda\to\infty} z(x,\lambda) = \frac{1}{u'}\exp\left[-\int_0^u \frac{h(v)-h(0)}{v}dv\right]\Phi(u), \tag{6.37}$$

where $\Phi(u)$ is the solution to Eq. (6.34) and u is given by the expression (6.32).

Moreover, we have obtained a uniform solution at the transition point x_0. Also, in the neighbourhood of x_0, $f(x) \approx f'(x_0)(x-x_0)$, and according to Eq. (6.32)

$$u \approx \left[\frac{3}{2}f(x_0)^{\frac{1}{2}}\int_{x_0}^{x}\sqrt{(x'-x_0)}\mathrm{d}x'\right]^{\frac{2}{3}} = f'(x_0)^{\frac{1}{3}}(x-x_0).$$

Hence $u'(x_0) = f'(x_0)^{\frac{1}{3}}$. Consequently, as $h(u) = g(x(u))/u'^3$, we obtain $h(0) = g(x_0)/f'(x_0) = \mu$.

These results lead naturally to the study of the comparison equation.

6.2.3 *The comparison equation*

The aim of this section is to give some analytic solutions to the equation [Langer (1955a,b)]

$$y''' - xy' - \mu y = 0. \tag{6.38}$$

A first remark about this equation is that a scaling of the x variable leads to the comparison equation (6.34) by putting $x = -\lambda^{2/3}\overline{x}$.

A solution to this equation can be obtained using the Laplace method [Ince (1956)]. Writing

$$y = \int_C \mathrm{e}^{zx} f(z)\mathrm{d}z,$$

with the condition

$$\left.\mathrm{e}^{-z^3/3}\right|_C = 0, \tag{6.39}$$

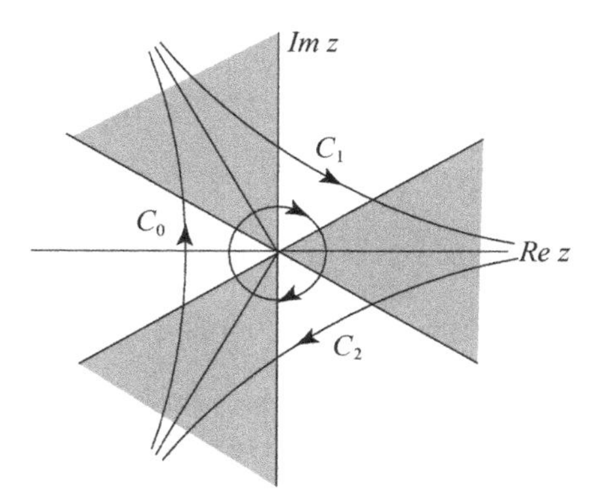

Fig. 6.2 Integration contour associated with the integral (6.40).

we obtain the integral representation

$$y = \int_{C_i} e^{-z^3/3+zx} z^{\mu-1} dz, \tag{6.40}$$

where the integration paths C_i are given on Fig. (6.2).

We now consider some particular cases of the comparison equation (6.38).

- The case where $\mu = 0$ leads to the Airy equation (2.1): $(y')'' - xy' = 0$, producing the general solution (cf. §2.1)

$$y_0 = a \int Ai(x) dx + b \int Bi(x) dx + c, \tag{6.41}$$

  where a, b, c are integration constants.
- The case where $\mu = 1$ corresponds to the inhomogeneous Airy equation (2.126) (cf. §2.3) $y'' - xy = c/\pi$, giving the solution[2]

$$y_1 = aAi(x) + bBi(x) + cHi(x). \tag{6.42}$$

- The case where $\mu = 1/2$ involves the square of Airy functions (cf. §2.4)

$$\begin{aligned} y_{1/2} = aAi^2\left(2^{-2/3}x\right) + bBi^2\left(2^{-2/3}x\right) \\ + cAi\left(2^{-2/3}x\right) Bi\left(2^{-2/3}x\right). \end{aligned} \tag{6.43}$$

  Note that this case is self-adjoint.

[2] It should be noted that if y is a solution of $y''' - xy' - \mu y = 0$, then $y' = z$ is a solution of $z''' - xz' - (\mu + 1)z = 0$.

- The case where $\mu = -1/2$ also involves the square of Airy functions

$$\begin{aligned} y_{-1/2} = a & \left[Ai'^2 \left(2^{-2/3} x \right) - 2^{-2/3} x \, Ai^2 \left(2^{-2/3} x \right) \right] \\ & + b \left[Bi'^2 \left(2^{-2/3} x \right) - 2^{-2/3} x \, Bi^2 \left(2^{-2/3} x \right) \right] \\ & + c \left[Ai' \left(2^{-2/3} x \right) Bi' \left(2^{-2/3} x \right) \right. \\ & \left. - 2^{-2/3} x \, Ai \left(2^{-2/3} x \right) Bi \left(2^{-2/3} x \right) \right] . \end{aligned} \tag{6.44}$$

Now we consider the more general case, where μ is any integer. We set $A(x) = aAi(x) + bBi(x)$, $A'(x)$ its derivative and $A_1(x)$ its primitive.[3] It can be seen (from the above results) that $y_0(x) = A_1(x)$. So if we set $z_n = y_{-n}$ for the parameter $\mu = -n$, $n \in \mathbb{N}$, we obtain the recurrence relation: $z'_n = n z_{n-1}$. Thus we can verify that the first solutions to Eq. (6.38) are

$$\begin{aligned} z_1 &= x A_1(x) - A'(x) \\ z_2 &= x^2 A_1(x) - A(x) - x A'(x) \\ z_3 &= (x^3 - 2) A_1(x) - x A(x) - x^2 A'(x). \end{aligned} \tag{6.45}$$

We can then generate the solutions to Eq. (6.38), thanks to the recurrence relation

$$z_n = x z_{n-1} - (n-1)(n-2) z_{n-3}, \tag{6.46}$$

allowing these solutions to be written explicitly as

$$z_n(x) = Pi_n(x) A_1(x) - Qi_{n-2}(x) A(x) - Ri_{n-1}(x) A'(x), \tag{6.47}$$

where the polynomials Pi_n, Qi_n and Ri_n obey the recurrence relations

$$Pi_n = x Pi_{n-1} - (n-1)(n-2) Pi_{n-3} \tag{6.48}$$

$$Qi_n = x Qi_{n-1} - n(n+1) Qi_{n-3} \tag{6.49}$$

$$Ri_n = x Ri_{n-1} - n(n-1) Ri_{n-3}. \tag{6.50}$$

The values of these polynomials[4] for $n = 0, 1, 2$ are respectively $1, x, x^2$.

It is interesting to note that, if the integration constants a and b cancel, the solutions are only given by the polynomial $z_n = Pi_n(x)$, which is simply the Airy polynomial (cf. §4.2.3). These solutions also allow us to obtain those corresponding to the adjoint equation, i.e. for $\mu = n+1$, $n \in \mathbb{N}$,

$$y''' - xy' - (n+1)y = 0. \tag{6.51}$$

[3]This primitive includes the additive integration constant c

[4]The properties of the polynomial Pi_n were already given in §4.2.3.

This equation has the following noteworthy recurrence properties

$$y_n = y'_{n-1}, \tag{6.52a}$$

$$y_n = xy_{n-1} - (n-2)y_{n-3}. \tag{6.52b}$$

According to the theorem given in §6.2.1, when we have the solutions for $\mu = -n$, we consequently obtain the solutions of Eq. (6.51). For example, in the case where $\mu = 0$, the solution

$$y_0(x) = a \int Ai(x)\mathrm{d}x + b \int Bi(x)\mathrm{d}x + c,$$

allows us to find the solution in the case where $\mu = 1$

$$y_1(x) = \alpha Ai(x) + \beta Bi(x) + \gamma Hi(x),$$

with $\alpha = ca' - c'a$, $\beta = c'b - cb'$, $\gamma = ab' - a'b$ and, thanks to the relation (2.130),

$$Hi(x) = Bi(x) \int_{-\infty}^{x} Ai(t)\mathrm{d}t - Ai(x) \int_{-\infty}^{x} Bi(t)\mathrm{d}t.$$

Another example is found from the first recurrence relation (6.52a), which allows the nth order derivative of an Airy function to be calculated. In fact, the first derivatives are Ai', $Ai'' = xAi$, $Ai''' = Ai + xAi'$, and for $n \geq 3$ one has

$$\begin{aligned} Ai^{(n)}(x) = \frac{1}{(n-1)!} & \left[\left(Pi_{n-1}Qi_{n-2} - Pi_nQi_{n-3} \right) Ai(x) \right. \\ & \left. + \left(Pi_{n-1}Ri_{n-1} - Pi_nRi_{n-2} \right) Ai'(x) \right]. \end{aligned} \tag{6.53}$$

With the same method, we can find the derivatives of the Scorer function $Hi(x)$ with the integral representation (for $n \geq 3$)

$$\begin{aligned} Hi^{(n)}(x) &= \frac{1}{\pi} \int_0^{\infty} \mathrm{e}^{-t^3/3+xt} t^n \mathrm{d}t \\ &= \frac{1}{(n-1)!} \left[\left(Pi_{n-1}Qi_{n-2} - Pi_nQi_{n-3} \right) Hi(x) \right. \\ &\quad + \left(Pi_{n-1}Ri_{n-1} - Pi_nRi_{n-2} \right) Hi'(x) \\ &\quad \left. + \frac{1}{\pi} \left(Qi_{n-2}Ri_{n-2} - Qi_{n-3}Ri_{n-1} \right) \right]. \end{aligned} \tag{6.54}$$

6.3 A differential equation of the fourth order

Langer used the comparison equation (6.38), to study the asymptotic solutions to the Orr–Sommerfeld equation (a fourth order differential equation), which describes the hydrodynamic instabilities of a Poiseuille flow. In this case the parameter λ^2 corresponds to the Reynolds number of the fluid. A more developed asymptotic method [Drazin & Reid (1981)] involves, as a comparison equation, a fourth order differential with two parameters

$$y'''' - xy'' - \alpha y' - \beta y = 0. \tag{6.55}$$

In the case where $\beta = 0$, we again find Eq. (6.38).

In the general case, the method of Laplace yields solutions to (6.55) with the following integral representations

$$y_k(\alpha, \beta; x) = \int_{C_k} e^{-z^3/3 + xz - \beta/z} z^{\alpha - 2} dz, \tag{6.56}$$

where the contours C_k are conveniently chosen (cf. §6.2.3). These solutions, which constitute generalisations of Airy functions, were studied in detail by Rabenstein (1958). We refer the reader to this article for further information. However, note that the adjoint equation of (6.55) may be written

$$z'''' - \frac{d^2}{dx^2}(xz) + \alpha z' - \beta z = 0, \tag{6.57}$$

or

$$z'''' - xz'' + (\alpha - 2)z' - \beta z = 0. \tag{6.58}$$

We see consequently the noteworthy result, for $\alpha = 1$, that Eq. (6.55) is self-adjoint. This result (6.58) allows us to generate the solutions of $y(2-\alpha, \beta; x)$ when the solutions $y(\alpha, \beta; x)$ with $\alpha > 0$ are known, as previously for Eq. (6.38).

A comment is in order on Eq. (6.38), as well as Eq. (6.55), concerning their Airy transform (see §4.2). In fact, this transform allows these equations to be rewritten in a much simpler form.

We start with the third order equation (6.38). Taking into account the properties of the Airy transform (§4.2), we have

$$\overline{\mathcal{A}}\,[xf] = x\varphi + \varphi''. \tag{6.59}$$

Hence, in this transform, Eq. (6.38) becomes a first order differential equation

$$x\frac{d\varphi}{dx} + \mu\varphi = 0. \tag{6.60}$$

A solution to equation (6.38) is then

$$y = \mathcal{A}\left[\varphi\right] = \int_{-\infty}^{+\infty} Ai(x - x') \frac{\mathrm{d}x'}{x'^{\mu}}. \tag{6.61}$$

We can easily recognise that this is the solution $Pi_n(x)$ to Eq. (6.38) in the case $\mu = -n$, namely the Airy polynomials. This method also works for the equation (6.55). The formula (6.59) gives

$$x\varphi'' + \alpha\varphi' + \beta\varphi = 0, \tag{6.62}$$

which is a Bessel equation of which the solutions for $\beta > 0$ are

$$\varphi = x^{(1-\alpha)/2} Z_{1-\alpha}\left(\frac{1}{2}\sqrt{\frac{x}{\beta}}\right), \tag{6.63}$$

where $Z_{1-\alpha}$ is a Bessel function which is chosen to ensure the convergence (and the continuity at $x = 0$) of the integral in the inverse Airy transform. A solution of (6.55) therefore reads

$$y = \int_{-\infty}^{+\infty} Ai(x - x') x'^{(1-\alpha)/2} Z_{1-\alpha}\left(\frac{1}{2}\sqrt{\frac{x'}{\beta}}\right) \mathrm{d}x'. \tag{6.64}$$

For instance, in the self-adjoint case $\alpha = 1$, one has J_0 for $x' > 0$ and I_0 for $x' < 0$.

We close this section on the generalisation of Airy functions, with the functions defined by the integrals [see Drazin & Reid (1981)]

$$A_k(p, q; x) = \frac{1}{2\mathrm{i}\pi} \int_{\mathrm{C}_k} \mathrm{e}^{-z^3/3 + xz} \left(\ln z\right)^q \frac{\mathrm{d}z}{z^p}. \tag{6.65}$$

They have noteworthy recurrence properties that were used recently in the search for analytic solutions to non-linear equations [Laurenzi (1993)].

Exercises

(1) Find three linearly independent solutions to the equations

$$y''' - xy' \pm \frac{3}{2}y = 0.$$

Hints: See footnote p. 121 and examples in §6.2.3, Eq. (6.44).

(2) Find a third order differential equation for the following functions

$$y(x) = \int_0^\infty Ai'(x+t)\frac{\mathrm{d}t}{\sqrt{t}},$$

$$z(x) = \int_0^\infty Ai_1(x+t)\frac{\mathrm{d}t}{\sqrt{t}}.$$

(3) Find a differential equation of the fourth order for which a solution is

$$y(x) = \int_0^\infty \cos\left(\frac{t^3}{12} + xt - \frac{\beta^2}{t} + \frac{\pi}{4}\right)\frac{\mathrm{d}t}{\sqrt{t}}.$$

Then, using a scaling, compare the result with Eq. (6.58). *Hint*: Express $y(x)$ in terms of Airy functions.

Chapter 7

Applications to Classical Physics

7.1 Optics and electromagnetism

Airy functions were introduced by G. B. Airy in 1838 in his article about the calculation of the light intensity in the neighbourhood of a caustic. We shall establish here the expression of this intensity, not in the way followed by Airy, but by the more "modern" approach of Landau & Lifchitz (1964).

Let us consider a monochromatic light source and an aperture in an opaque screen. According to the laws of geometrical optics, beyond this screen, space is separated into two zones; the dark zone presenting a clear border with the illuminated zone. However, the longer the wavelength of the source compared to the dimension of the aperture, the more intense the phenomenon of diffraction, thus complicating the distribution of the light intensity in the neighbourhood of this border. According to the Huygens principle, we consider each element of the surface $\mathrm{d}S$ of the aperture to be the source of a spherical wave. That is to say, u being the amplitude of the field on $\mathrm{d}S$ and k the wave number of the source of light, the electromagnetic field in a point P located at a distance R from the aperture, is proportional to the sum of these spherical waves

$$u_P \propto \int u \frac{\mathrm{e}^{\mathrm{i}kR}}{R} \mathrm{d}S_\mathrm{n}, \tag{7.1}$$

where $\mathrm{d}S_\mathrm{n}$ is the projection of $\mathrm{d}S$ on the normal plane, corresponding to the direction of the ray resulting from the source, and arriving on the surface of the aperture (cf. Fig. (7.1)). If moreover we consider that u is constant on the surface of the aperture, the field in P is

$$u_P \propto \int \frac{\mathrm{e}^{\mathrm{i}kR}}{R} \mathrm{d}S_\mathrm{n}. \tag{7.2}$$

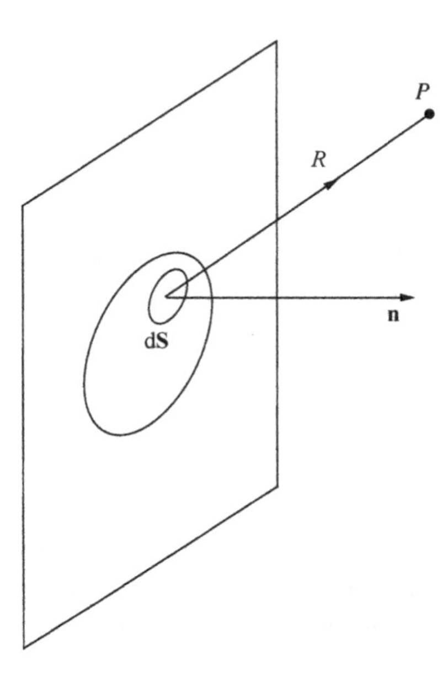

Fig. 7.1 Diffraction by an aperture in an opaque screen.

Let us consider now the particular case where the point P is located in the neighbourhood of a caustic, which is the line separating the shaded zone from the illuminated zone, in the ideal case of geometrical optics. We note that the caustic is also defined, in geometrical optics, as being the surface where the light intensity becomes infinite.

First, we consider only one section $\mathscr{C}$ in the plane (x, y) of the surface defined previously. The caustic $\mathscr{C}'$ is then the envelope of $\mathscr{C}$. We seek to establish the light intensity in the neighbourhood of the point O, which is the contact point between the ray $M'O$ and the caustic. D indicates the length of the segment $[M'O]$. P is the point located in the neighbourhood of O where we shall calculate the intensity of the electromagnetic field: this point is located by its ordinate y (cf. Fig. (7.2)). We denote C the curvature center of the caustic and ρ the radius of curvature.

The variable R in the expression (7.2) is then the distance from a point M of the wave surface $\mathscr{C}$, from the point P (see Fig. (7.2)). The geometrical properties of the envelope make it possible to establish the equality of the angles (M, H, M') and (O, C, O'), which is denoted α. If we assume that we are located at a point P far from the wave surface, i.e. α is small enough to be expanded up to the third order: $\sin\alpha \approx \theta - \theta^3/6$, and that the points

O and O' are sufficiently close to consider that the radius of curvature ρ is constant on the arc OO', the following relations can be established

$$R = MP \approx MO - y \sin\alpha \quad .$$
$$MO \approx MO' + \rho \sin\alpha$$
$$MO' \approx M'O' - \alpha\rho.$$

Thus we obtain

$$R \approx D - y\alpha - \frac{\rho\alpha^3}{6}, \tag{7.3}$$

and the formula (7.2) is reduced to

$$u_P \propto \int \frac{\mathrm{e}^{\mathrm{i}kR}}{R} \mathrm{d}\alpha.$$

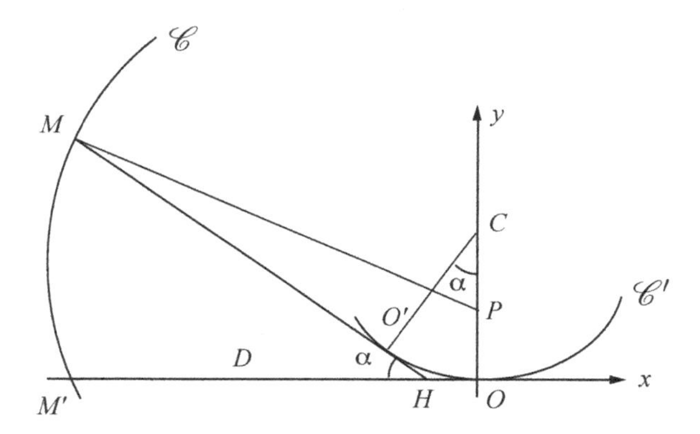

Fig. 7.2 Calculation of the light intensity at the point P in the neighbourhood of the caustic $\mathscr{C}'$.

As the factor $1/R$ varies slowly according to α, this variation will be neglected compared to the exponential $\frac{1}{R} \approx \frac{1}{D}$ (this is the principle of the stationary phase method, cf. §5.1). We obtain, finally, the expression of the amplitude at the point P

$$u_P \propto \int \mathrm{e}^{-\mathrm{i}k\left(y\alpha + \rho\alpha^3/6\right)} \mathrm{d}\alpha.$$

Note that we find here the expression given by Hochstadt (1973), which carries out the same type of calculation by considering plane waves reflected by a concave surface. By applying the change of variable $t = \left(\frac{k\rho}{2}\right)^{1/3} \alpha$, this expression becomes

$$u_P \propto \int \mathrm{e}^{\mathrm{i}\left(t^3/3 + \left(2k^2/\rho\right)^{1/3} yt\right)} \mathrm{d}t.$$

We recognise the integral definition (2.21) of the Airy function Ai. The amplitude can thus be written

$$u_P \propto Ai\left[\left(\frac{2k^2}{\rho}\right)^{1/3} y\right].$$

By reintroducing a proportionality factor, the light intensity in the vicinity of a caustic becomes

$$I_P = A\ Ai^2\left[\left(\frac{2k^2}{\rho}\right)^{1/3} y\right]. \tag{7.4}$$

It is noticeable, in particular, that the intensity is not maximum on the caustic, but in the illuminated zone, at a position determined by the first maximum of the function Ai: $\left(\frac{2k^2}{\rho}\right)^{1/3} y \simeq -1.02$. For "large" values of y, the expression (7.4) becomes (cf. §2.1.4.3)

$$I_P = \frac{1}{\sqrt{y}} \exp\left[-\frac{4}{3}\left(2k^2/\rho\right)^{1/2} y^{3/2}\right]. \tag{7.5}$$

In the shaded zone, the intensity decreases exponentially. For the negative values of y, the formula (7.4) gives the asymptotic expression

$$I_P \propto \frac{1}{\sqrt{-y}} \cos^2\left[\frac{2}{3}\left(\frac{2k^2}{\rho}\right)^{1/2} (-y)^{3/2} - \frac{\pi}{4}\right]. \tag{7.6}$$

In the illuminated zone, the intensity oscillates quickly according to y. Its average value is

$$\langle I_P \rangle \propto \frac{1}{\sqrt{-y}},$$

i.e. we find here the intensity given by geometrical optics.

7.2 Fluid mechanics

7.2.1 *The Tricomi equation*

The stationary, two-dimensional flow (in a (x, y) plane) of a compressible gas obeys the Tchaplyguine equation [Landau & Lifchitz (1971)]

$$\frac{\partial^2 \Phi}{\partial \theta^2} + \frac{v^2}{1 - v^2/c^2} \frac{\partial^2 \Phi}{\partial v^2} + v \frac{\partial \Phi}{\partial v} = 0, \tag{7.7}$$

where v is the speed of gas, θ the angle between $\vec{v}$ and the x axis, and c the speed of sound in gas. Φ is a function of speed defined by

$$\Phi = -\phi + x v_x + y v_y, \tag{7.8}$$

where ϕ is the velocity potential $\vec{v} = \vec{\nabla}\phi$. The relation (7.8), in fact, defines a Legendre transformation, which brings the non-linear equation of motion back to the linear Eq. (7.7). The price to be paid for this linearisation is that the boundary conditions become non-linear [Landau & Lifchitz (1971)].

At the transonic limit, the velocity of gas approaching the speed of sound, the third term in Eq. (7.7) becomes negligible in comparison to the second term. Moreover, without going into the details of this system, we can introduce a new variable η, known as the "variable of Tchaplyguine" [Hayasi (1971)], which depends on the velocity v (as well as the critical speed of sound and other parameters), such that Eq. (7.7) becomes

$$\frac{\partial^2 \Phi}{\partial \eta^2} - \eta \frac{\partial^2 \Phi}{\partial \theta^2} = 0. \tag{7.9}$$

This partial derivative equation is called the Tricomi equation or Euler–Tricomi. Gramtcheff pursues a mathematical study of this equation and its relationship to the Airy equation [Gramtcheff (1981)].

Note first of all that, for $\eta > 0$, Eq. (7.9) is a hyperbolic equation and for $\eta < 0$, an elliptic one.

Then consider the Fourier transform: $\tilde{\Phi}(\eta, \omega)$ of the function $\Phi(\eta, \theta)$

$$\tilde{\Phi}(\eta, \omega) = \int \mathrm{e}^{-\mathrm{i}\omega\theta} \Phi(\eta, \theta) \mathrm{d}\theta.$$

By applying this transformation to Eq. (7.9), we obtain

$$\frac{\partial^2 \tilde{\Phi}}{\partial \eta^2} + \eta \omega^2 \tilde{\Phi} = 0.$$

Through the change of variable $\xi = \eta \omega^{2/3}$, we then find the Airy equation

$$\frac{\partial^2 \tilde{\Phi}}{\partial \xi^2} + \xi \tilde{\Phi} = 0. \tag{7.10}$$

The solution to this equation being of the type $\tilde{\Phi} = Ai(-\xi) = Ai(-\eta\omega^{2/3})$. The solution to the Tricomi equation, is determined by the inverse Fourier transform

$$\Phi(\eta, \theta) = \int \mathrm{e}^{\mathrm{i}\omega\theta} Ai(-\eta\omega^{2/3}) \tilde{f}(\omega) \mathrm{d}\omega,$$

where $\tilde{f}(\omega)$ is the Fourier transform of the initial angular profile.[1]

In fact, the general integral of the Tricomi equation can be written

$$\Phi(\eta, \theta) = \int_{\mathrm{C}} g\left(\frac{z^3}{3} - \eta z + \theta\right) \mathrm{d}z, \tag{7.11}$$

[1]Note that the separation of variables method leads to the same kind of solutions.

where g is an arbitrary function such that g' takes the same values at the ends of the integration path C. Assuming $\tilde{g}$ is the Fourier transform of g, the relation (7.11) becomes

$$\Phi(\eta,\theta)=\int_{C}\int_{-\infty}^{+\infty}\mathrm{e}^{\mathrm{i}(z^3/3-\eta z+\theta)t}\tilde{g}(t)\mathrm{d}t\mathrm{d}z.$$

By gathering the terms in z, this last relation can also be written

$$\Phi(\eta,\theta)=\int_{-\infty}^{+\infty}\mathrm{d}t\ \mathrm{e}^{\mathrm{i}\theta t}\tilde{g}(t)\int_{C}\mathrm{d}z\ \mathrm{e}^{\mathrm{i}(z^3/3-\eta z)t}. \tag{7.12}$$

Under the condition of choosing a convenient integration path C, we find here the integral expression of the Airy function Ai.

7.2.2 *The Orr–Sommerfeld equation*

7.2.2.1 *Plane flow of an incompressible viscous fluid*

The Navier equation for an incompressible viscous fluid comes from the general equations of motion of a fluid

$$\begin{cases}\frac{\partial\vec{u}}{\partial t}+\vec{u}.\vec{\nabla}\vec{u}+\vec{\nabla}\left(\frac{p}{\rho}\right)=\nu\Delta\vec{u}+\frac{\vec{f}}{\rho}\\ \vec{\nabla}.\vec{u}=0,\end{cases}$$

where the density $\rho=\rho_0$ is a constant. We will also assume that the external force $\vec{f}$ comes from a potential. By introducing the vorticity

$$\vec{\omega}=\frac{1}{2}\vec{\nabla}\wedge\vec{u},$$

and noticing that

$$\vec{u}.\vec{\nabla}\vec{u}=2\vec{\omega}\wedge\vec{u}+\frac{1}{2}\vec{\nabla}\left|\vec{u}\right|^2,$$

we can transform the Navier equation into the vorticity equation

$$\frac{\partial\vec{\omega}}{\partial t}+\vec{\nabla}\wedge(\vec{\omega}\wedge\vec{u})=\nu\Delta\vec{\omega}. \tag{7.13}$$

In the particular case of the plane flow $\vec{\omega}=\omega\vec{k}$, the velocity field can be obtained from the current function $\psi(x,y,t)$ by the relations

$$\begin{cases}u_x=\frac{\partial\psi}{\partial y}\\ u_y=-\frac{\partial\psi}{\partial x}.\end{cases}$$

For the vorticity component, we obtain

$$\begin{aligned}\omega &= \frac{1}{2}\left(\frac{\partial u_y}{\partial x} - \frac{\partial u_x}{\partial y}\right) \\ &= -\frac{1}{2}\left(\frac{\partial^2 \psi}{\partial x^2} + \frac{\partial^2 \psi}{\partial y^2}\right) \\ &= -\frac{1}{2}\Delta\psi.\end{aligned} \tag{7.14}$$

Taking the curl of the first member of Eq. (7.13) yields

$$\vec{\nabla} \wedge (\vec{\omega} \wedge \vec{u}) = \left(\frac{\partial \omega}{\partial x}\frac{\partial \psi}{\partial y} - \frac{\partial \omega}{\partial y}\frac{\partial \psi}{\partial x}\right)\vec{k}. \tag{7.15}$$

Consequently, taking into account Eqs. (7.14) and (7.15), the vorticity equation is reduced to the quasilinear, fourth order equation for the current function:

$$\left[\frac{\partial}{\partial t} + \frac{\partial \psi}{\partial y}\frac{\partial}{\partial x} - \frac{\partial \psi}{\partial x}\frac{\partial}{\partial y} - \frac{1}{\mathrm{Re}}\Delta\right]\Delta\psi = 0. \tag{7.16}$$

In this equation, we scale the variables so that the Reynolds number Re appears instead of the viscosity.

Equation (7.16), in the extreme cases Re $= 0$ and Re $= \infty$, leads to remarkable equations for the fluid. For instance, the case Re $= 0$ produces the biharmonic equation

$$\Delta^2\psi = 0,$$

which presents the disadvantage (in the nonstationary case) of leading to an ill-posed problem. For the case Re $= \infty$, it gives the Euler equation (the stationary case)

$$\frac{\partial \psi}{\partial y}\frac{\partial \Delta\psi}{\partial x} - \frac{\partial \psi}{\partial x}\frac{\partial \Delta\psi}{\partial y} = 0$$

leading to the resolution of

$$\Delta\psi = F(\psi),$$

where F is an arbitrary function of the current ψ. The case Re $\gg 1$ leads to the Prandtl equation for the boundary layer, but also to the study of the stability of a plane flow and to the Orr–Sommerfeld equation.

7.2.2.2 *Stability of an almost parallel flow*

In this section, we consider the perturbation of the velocity field along the x-axis, parallel to the flow. We thus assume that we have, upstream,

$$\begin{cases} u_x = U(y) = \frac{\partial \psi_0}{\partial y} \neq 0 \\ u_x = 0. \end{cases}$$

The velocity field $U(y)$ should be specified as being strictly parallel. In particular, the pressure scalar field should depend only on x, and the pressure gradient $\frac{\mathrm{d}p}{\mathrm{d}x}$ should be constant, because of the equation of motion:

$$\nu \frac{\mathrm{d}^2 U}{\mathrm{d}y^2} = \frac{1}{\rho} \frac{\mathrm{d}p}{\mathrm{d}x} = \text{constant.}$$

Thus $U(y)$ should be a quadratic function of y. Then two particular cases emerge:

- the Couette flow where $U(y) = y$,
- the Poiseuille flow where $U(y) = 1 - y^2$,

written by conveniently standardising the variables. Hence, we seek the perturbed solution $\psi_1(t, x, y)$ by writing that

$$\psi(t, x, y) = \psi_0(y) + \eta \psi_1(t, x, y),$$

is the solution to Eq. (7.16), where η is a "small" parameter. We thus find ψ_1, by keeping only the dominant terms according to η, satisfying the linear equation

$$\frac{\partial \Delta \psi_1}{\partial t} + U \frac{\partial \Delta \psi_1}{\partial x} - U'' \frac{\partial \psi_1}{\partial x} - \frac{1}{\mathrm{Re}} \Delta^2 \psi_1 = 0. \tag{7.17}$$

Because we are seeking a perturbation solution of the speed distribution $U(y)$, it is natural to look for a solution ψ_1 in the form of a wave that propagates along the x axis with a speed c:

$$\psi_1(t, x, y) = \phi(y)\, \mathrm{e}^{\mathrm{i}\alpha(x - ct)}. \tag{7.18}$$

By plugging Eq. (7.18) into the perturbed equation, we obtain

$$\frac{1}{\mathrm{i}\alpha \mathrm{Re}} \left[\frac{\mathrm{d}^2}{\mathrm{d}y^2} - \alpha^2 \right]^2 \phi + (c - U) \left[\frac{\mathrm{d}^2}{\mathrm{d}y^2} - \alpha^2 \right] \phi + U'' \phi = 0, \tag{7.19}$$

which is nothing else but the Orr–Sommerfeld equation. It is necessary to add to this equation the boundary conditions $\phi(y) = \phi'(y) = 0$ when $y = y_1$ or $y = y_2$. Thus, to obtain a non-trivial solution to this problem, it is necessary to find a relation between the various parameters

$$\mathcal{S}(\alpha, c; \mathrm{Re}) = 0,$$

with this relation defining the eigenvalues of the problem. The reader will find a detailed analysis of the Orr–Sommerfeld equation in the work by Drazin & Reid (1981).

Among the various cases in which Airy functions are involved as solutions to the Orr–Sommerfeld equation (7.19), we shall consider only two.

We first consider the case of a Couette flow. We have $U(y) = y$, and thus $U''(y) = 0$. The equation is simplified by writing

$$\left(\frac{\mathrm{d}^2}{\mathrm{d}z^2} - \alpha^2\right)\phi = \psi. \tag{7.20}$$

So we obtain

$$\varepsilon^2\left(\frac{\mathrm{d}^2}{\mathrm{d}z^2} - \alpha^2\right)\psi - z\psi = 0, \tag{7.21}$$

where we note $z = y - c$ and $\varepsilon = (\mathrm{i}\alpha\mathrm{Re})^{-1/3}$. The boundary conditions are $\phi' = \frac{\mathrm{d}\phi}{\mathrm{d}z} = 0$ at $z = \pm 1 - c$. The solution to Eq. (7.21) is expressed in terms of Airy functions and the solution to Eq. (7.20) can then be obtained by using the method of variation of parameters [Drazin & Reid (1981)].

For the second case, we consider the case of large Reynolds numbers, or more precisely, we look for asymptotic solutions according to the parameter

$$\varepsilon^3 = \frac{1}{\mathrm{i}\alpha\mathrm{Re}} \ll 1.$$

The general analysis of the problem was carried out by Langer (1955a,b) as pointed out in §6.2, but also by Rabenstein (1958) (as well as by other authors cited in Drazin & Reid (1981)). We shall just mention the method that consists of taking, as starting equation, the Orr–Sommerfeld equation which is "truncated", removing its less important terms

$$\frac{1}{\mathrm{i}\alpha\mathrm{Re}}\phi'''' + (c - U)\phi'' = 0. \tag{7.22}$$

We are then in a case where the uniform approximation method for a differential equation can be applied (cf. §5.2). Indeed, as the unspecified velocity field $U(y)$ has a turning point $U(y_t) = c$, the solution expressed by this method will be uniformly valid in the neighbourhood of this turning point.

7.3 Elasticity

Let us consider the system formed by a homogeneous rod with a small circular cross-section, which is in equilibrium. This rod is a one dimensional system, because its cross-section Σ_p is comparable to a geometrical point P, located by its curvilinear coordinate s. The vectorial equations of equilibrium are [Landau & Lifchitz (1967)]

$$\frac{\mathrm{d}\vec{T}}{\mathrm{d}s} + \vec{f} = 0 \tag{7.23}$$

$$\frac{\mathrm{d}\vec{M}}{\mathrm{d}s} + \vec{u} \wedge \vec{T} + \vec{m} = 0, \tag{7.24}$$

where $\vec{u}$ is a normal unit vector to the section, and directed towards the increasing values of s, where $\vec{T}(s)$ and $\vec{M}(s)$ are the reduced elements of the torque (i.e. the resultant and the moment of the torque) of the interior constraints in P, and where $\vec{f}(s)$ and $\vec{m}(s)$ are the reduced elements of the torque of external constraints.

In the case we are studying i.e. weak inflexion without torsion of a rod, we have

$$\begin{cases} \vec{m}(s) = 0 \\ T_T = 0 \\ M_u = 0 \end{cases}$$

where $\vec{m}$ is the linear distribution of torques, $\vec{T}$ is the normal component to $\vec{u}$ (there is no shearing stress), and $M_u = \vec{M}.\vec{u}$ is the torsion moment. The interior and external constraints are thus reduced to the tension (or normal compression) $T = T_u = \vec{T}.\vec{u}$ and the bending moment (normal component with $\vec{u}$) $M = M_F$.

A rod of length L is placed vertically in the gravity field. This rod is free at the top and embedded, at the bottom, in a slab of concrete (cf. Fig. (7.3)). We choose the referential (x, y, z) so that deformation takes place in the (x, z) plane; $s(x, y, z)$ is then reduced to $s(x, 0, z)$. Moreover we consider a "weak" flexion (the radius of curvature of the rod is at any point much smaller than its length), so that the vector $\vec{u}$ is directed along the z axis and that its norm depends only on x.

If we denote I the moment of inertia of the rod and E its Young modulus (the quantity IE is the rigidity flexion of the rod), the flexion momentum is written [Landau & Lifchitz (1967)]

$$\vec{M} = (0, IEx'', 0).$$

Then Eq. (7.24) gives the relation:

$$IEx''' = T_x x'. \tag{7.25}$$

Equation (7.23) allows us to determine T_x, q being the weight per unit length of the rod. We obtain, under the condition $T(L) = 0$ (the top of the rod being free),

$$T_x = -q(L - z). \tag{7.26}$$

The equilibrium equation of the rod is thus

$$IEx''' = -q(L - z)x'. \tag{7.27}$$

By writing $u = x'$ and $\eta = \left(\frac{q}{EI}\right)^{1/3}(z - L)$, we obtain the Airy equation (2.1) $u'' = \eta u$, with the solution $u = aAi(\eta) + bBi(\eta)$.

Hence we obtain the form of the arrow, x, of the rod (its horizontal variation compared to the vertical position), according to the primitives of Ai and Bi. The constants a and b are determined from some limit conditions. In particular, at the upper end of the rod, we know that the bending moment is zero: $\left.\frac{\mathrm{d}^2x}{\mathrm{d}z^2}\right|_{z=L} = 0$. Since the lower end of the rod is fixed, we also have $\left.\frac{\mathrm{d}x}{\mathrm{d}z}\right|_{z=0} = 0$. These relations determine the critical length L of the rod as a function of the parameters of the material $L = 1.986\left(IE/q\right)^{1/3}$.

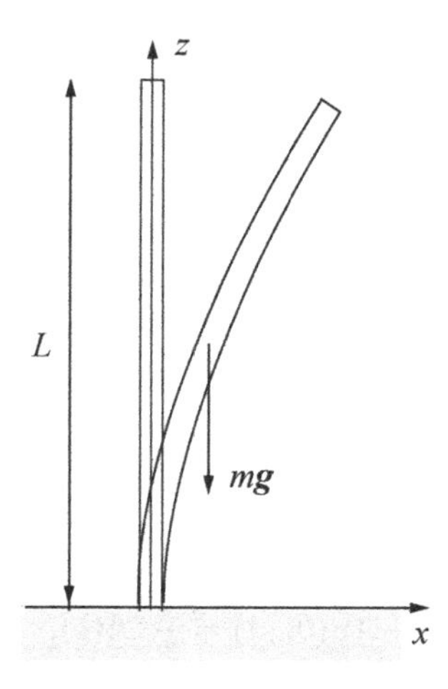

Fig. 7.3 Flexion of a rod placed vertically in the gravity field.

7.4 The heat equation

In this section, we shall see that the heat equation

$$\partial_t u = \partial_{xx} u, \tag{7.28}$$

may have a solution which can be expressed in terms of Airy functions. In fact, a Lie group analysis of this equation [Olver (1998)] produces the following solution

$$u(x,t) = \exp(2t^3/3 - xt)\, Ai(t^2 - x), \quad t \geq 0. \tag{7.29}$$

This result alternatively can be found by using the general solution to the heat equation as the convolution integral

$$u(x,t) = \frac{1}{\sqrt{4\pi\nu t}} \int_{-\infty}^{+\infty} \psi(y) \exp\left[-\frac{(x-y)^2}{4\nu t}\right] \mathrm{d}y, \tag{7.30}$$

where the initial condition $\psi(x)$ is chosen as $\psi(x) = Ai(x)$. In other words, this solution is the Airy transform of the Gaussian (see Eq. (4.42))

$$\frac{1}{\sqrt{4\pi t}} \exp\left[-\frac{x^2}{4t}\right],$$

leading to Eq. (7.29).

Surprisingly, Eq. (7.29) appears in a number of very different scientific domains. In addition to the case of the heat equation itself, Eq. (7.29) is also used in probability calculus. A considerable number of asymptotic distributions, arising in random combinatorics and analysis of algorithms, are of the exponential-quadratic type, i.e. Gaussian. However when confluences of critical points and singularities occur, the Airy function immediately appears. Recently, a new probability distribution, called the "map–Airy distribution" was introduced [Banderier *et al.* (2000)]. This distribution concerns the statistical properties of random maps, i.e. the question of what such random maps typically look like. It is defined by the probability density[2]

$$\mathcal{A}(s) = -2\exp(2s^3/3)\left[sAi(s^2) + Ai'(s^2)\right], \quad s \in \mathbb{R}. \tag{7.31}$$

In the solution (7.29), if we set $x = 0$, we have

$$\frac{\mathrm{d}}{\mathrm{d}s}\left(u(0,s)\right) = -s\mathcal{A}(s).$$

Moreover, we have

$$\frac{\mathrm{d}}{\mathrm{d}s}\left(\exp(2s^3/3)\, Ai'(s^2)\right) = -s^2\mathcal{A}(s),$$

and the following integral representation

$$\mathcal{A}(s) = \sqrt{\frac{3}{4\pi}} \int_0^{\infty} \mathrm{e}^{-u^3/12}\, Ai(us)\, u^{3/2}\, \mathrm{d}u\,. \tag{7.32}$$

Figure (7.4) plots the density $\mathcal{A}(s)$ where it can be seen that it decreases slowly when $s \to \infty$, as $s^{-5/2}$, and increases rapidly for the negative values as $\exp(s^3)$.

[2]We have change $s \to -s$ in contrast to the work of Banderier *et al.*

Another intriguing result related to the solution of the heat equation (7.29) appeared in research on the rational solutions to the Painlevé equation. The result is that the function $\theta(x,t) = \exp(2t^3/3)\, Ai(t^2 - x)$ gives rise to an asymptotic expansion

$$\frac{\partial}{\partial t} \ln \theta(x,t) \sim \sum_{n=0}^{\infty} a_n(x)\,(-2t)^{-n},$$

of which the coefficients $a_n(x)$ are involved in the rational solution to the Painlevé equation [Iwasaki *et al.* (2002)]. For further details, we refer the reader to this paper. The Painlevé equation, in relation with Airy functions, will be discussed in the next section.

7.5 Nonlinear physics

7.5.1 *Korteweg–de Vries equation*

The history of solitons started around 1838, when John Scott Russell, a Scottish engineer, observed a phenomenon on a canal close to Edinburgh which astounded him. A barge drawn along the canal suddenly stopped, but the bow wave continued its course along the canal. Russell followed it with his horse for several kilometers. This wave was propagating identical to itself, at a constant speed, as opposed to what we are used to seeing. Russell performed many experiments, from which he deduced that the propagation

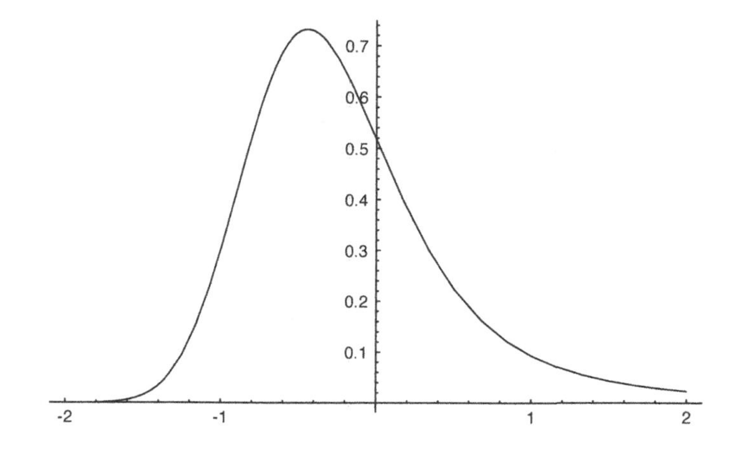

Fig. 7.4 The probability density $\mathcal{A}(s)$.

velocity of the wave is dependent on its amplitude η:

$$v = \sqrt{g(h+\eta)},$$

where g is the gravitational constant and h the depth of the canal. In 1844, he published a report of his observations.

Some time later, in 1845, G. B. Airy published a report on "tides and waves". He established (by theory) the existence of waves propagating with a speed close to that already determined by Russell's experiments. Airy concluded, however, that "solitary waves" do not exist, even though theory described the experiments correctly. A controversy then blew up between the two men, not dying out until three years after the death of Airy. In 1895, Korteweg and his student de Vries deduced an equation describing solitary waves in shallow water. The irony of the story, however, is that the solutions to the Korteweg–de Vries equation are closely related to Airy functions!

7.5.1.1 *The linearised Korteweg–de Vries equation*

The Korteweg–de Vries equation [Ablowitz & Segur (1981)]

$$\frac{\partial f}{\partial t} - 6f\frac{\partial f}{\partial x} + \frac{\partial^3 f}{\partial x^3} = 0, \tag{7.33}$$

and the modified Korteweg–de Vries equation

$$\frac{\partial f}{\partial t} - 6f^2\frac{\partial f}{\partial x} + \frac{\partial^3 f}{\partial x^3} = 0, \tag{7.34}$$

can both be linearized, in terms of the amplitudes, giving the equation

$$\frac{\partial f}{\partial t} + \frac{\partial^3 f}{\partial x^3} = 0, \tag{7.35}$$

which is the starting equation of the work by Widder (1979). Indeed an Airy transform upon the variable x (t being regarded as a parameter), yields the equation (see Lemma 4.9)

$$\frac{\partial \varphi_a}{\partial t} + \left(1 + \dot{a}a^2\right)\frac{\partial^3 \varphi_a}{\partial x^3} = 0.$$

We set $a = -(3t)^{1/3}$ so that the term with the third order derivative disappears and the solution of the transformed equation is an arbitrary function of the space (i.e. the initial condition). The inverse transform then leads to the solution

$$f(x,t) = \frac{1}{(3t)^{1/3}} \int_{-\infty}^{+\infty} Ai\left(\frac{x-y}{(3t)^{1/3}}\right)\phi(y)\mathrm{d}y. \tag{7.36}$$

This solution is alternatively obtained by carrying out a Fourier transform on Eq. (7.35) as proposed by Ablowitz & Segur (1981).

The solution (7.36) does not however represent a solitary wave, since the amplitude is a decreasing function of time. It is the competition between the dispersive term $\partial^3 f/\partial x^3$ and the non-linear term, which generates the soliton. To create a soliton, it is necessary to look for a solution as a wave $f(x-ct)$, travelling with the constant speed c. In this case, the Korteweg–de Vries equation (7.33) is equivalent to the ordinary differential equation

$$f''' - 6ff'' - cf' = 0.$$

If we look for the solutions of this equation under the condition that f, and its successive derivatives f' and f'', decrease towards zero when $z = x - ct$ tends towards infinity, we obtain

$$f(x - ct) = f(z) = -\frac{c}{2}\frac{1}{\mathrm{ch}^2\left(\frac{\sqrt{c}}{2}(z - z_0)\right)}. \tag{7.37}$$

Except for the sign (which can be restored by changing the sign of the non-linear term), this solution corresponds to the observation by Russell: a solitary wave whose propagation velocity is as large as the amplitude.

7.5.1.2 *Similarity solutions*

The search for a similarity group for a differential equation leads to what is called a similarity solution. If for the variables x and t, and the function f, we carry out the transformations ($a \in \mathbb{R}^+$)

$$\begin{cases} x = a^{\chi}\overline{x} \\ t = a^{\tau}\overline{t} \\ f = a^{\varphi}\overline{f}, \end{cases}$$

then the Korteweg–de Vries equation is invariant, should the following relations hold

$$\begin{cases} \varphi = \chi - \tau \\ \tau = 3\chi. \end{cases}$$

It can be seen therefore, that the variables $z = x(3t)^{-1/3}$ and $\phi = (3t)^{2/3} f(z)$, are also invariant in this similarity group. A similarity solution to Eq. (7.33) can then be written

$$f(x,t) = \frac{1}{(3t)^{2/3}}\,\phi\left(\frac{x}{(3t)^{1/3}}\right),$$

where the function ϕ is the solution to the ordinary differential equation

$$\phi''' - 6\phi\phi' - z\phi' - 2\phi = 0.$$

For low amplitudes of ϕ, linearisation leads to a solution in terms of the derivative of the Airy function, Ai'

$$f_L(x,t) = \frac{1}{(3t)^{2/3}} \, Ai'\left(\frac{x}{(3t)^{1/3}}\right).$$

It is noteworthy that the linearised similarity solution has the same similarity group as that of the linearised solutions (7.36) [Ablowitz & Segur (1981)].

The same method can be applied to the modified Korteweg–de Vries equation (7.34). The similarity group is given by

$$\begin{cases} 2\varphi = \chi - \tau \\ \tau = 3\chi. \end{cases}$$

The invariants are now given by the variables $z = x(3t)^{-1/3}$ and $\phi = (3t)^{1/3} f(z)$. The similarity solution to the modified Korteweg–de Vries equation is then written

$$f(x,t) = \frac{1}{(3t)^{1/3}} \, \phi\left(\frac{x}{(3t)^{1/3}}\right),$$

where ϕ is the solution to the ordinary differential equation

$$\phi''' - 6\phi^2\phi' - z\phi' - \phi = 0. \tag{7.38}$$

Among the linearised solutions of this equation, we find the solution

$$f_L(x,t) = \frac{1}{(3t)^{1/3}} \, Ai\left(\frac{x}{(3t)^{1/3}}\right),$$

which belongs to the family defined by Eq. (7.36).

Let us consider now Eq. (7.38). An integration leads to

$$\phi'' - 2\phi^3 - z\phi + a = 0, \tag{7.39}$$

where a is an integration constant. Equation (7.39) is called the "second Painlevé equation", or $P_{\rm II}$. This equation is treated in the following section. For further details on the theory of solitons and on solutions to the Korteweg–de Vries equation, we refer the reader to the book by Ablowitz & Segur (1981), see also Ablowitz & Clarkson (1991).

7.5.2 *The second Painlevé equation*

7.5.2.1 *The Painlevé equations*

At the end of the 19th century and at the beginning of the 20th century, mathematicians started to classify differential equations [Ince (1956)]. For this purpose they considered the class of rational equations in the unknown function of which only the singular points, depending on the initial conditions, are poles, and of which the critical points (branching points, essential singularities ...) are fixed. By definition therefore, the solutions are said to satisfy the Painlevé property. Thus Fuchs demonstrated in 1884 that, for the first order equation,

$$\frac{\mathrm{d}y}{\mathrm{d}z} = F(y, z),$$

where F is rational in y, and at least locally analytical in z. The only equation which does not have a moving critical point (i.e. a critical point that depends on the initial conditions) is the Riccati equation:

$$y' = P(z)y^2 + Q(z)y + R(z).$$

Painlevé and his co-workers (Gambier, Boutroux) were interested in the second order equation

$$\frac{\mathrm{d}^2y}{\mathrm{d}z^2} = F(y', y, z).$$

These mathematicians found fifty canonical equations, rational in y and y', and locally analytical in z, satisfying the Painlevé property. Six of these equations cannot be solved in terms of known functions; we refer to these equations by the name of "Painlevé transcendents".

The second Painlevé equation P_{II} is the object of the present section

$$y'' = zy + 2y^3 + a, \tag{7.40}$$

where a is a parameter. This equation can be considered as a *non-linear generalisation of the Airy equation.* Indeed, the solutions to Eq. (7.40) are closely related to the Airy functions, as we shall now see.

The case $a = 1/2$ is remarkable, in the sense that a solution to Eq. (7.40) can be expressed simply with the Airy functions. We first consider the Riccati equation

$$y' = y^2 + \frac{z}{2}. \tag{7.41}$$

If we assume $y = -w'/w$, this equation becomes

$$w'' + \frac{z}{2}w = 0, \tag{7.42}$$

which is nothing but an Airy equation. By taking the derivative of Eq. (7.41), we obtain

$$y'' = 2yy' + \frac{1}{2}.$$

Then we can replace y' by its value coming from Eq. (7.41) to obtain Eq. (7.40)

$$y'' = 2y\left(y^2 + \frac{z}{2}\right) + \frac{1}{2}.$$

The Painlevé equation has the following interesting property that, if $y(x;a)$ is a solution of P_{II}, then

$$\tilde{y}(x; a+1) = -y(x;a) - \frac{1+2a}{2y^2 + 2y' + x} \tag{7.43}$$

is also a solution of P_{II}. Therefore, we can deduce a solution for the case $a = n + \frac{1}{2}$ successively from the case $a = \frac{1}{2}$. For example, we have [Ablowitz & Clarkson (1991)]

$$y(x;\frac{3}{2}) = \frac{2(w'/w)^3 + x(w'/w) - 1}{2(w'/w)^2 + x}, \tag{7.44}$$

$$y(x;\frac{5}{2}) = \frac{4x(w'/w)^4 + 6(w'/w)^3 + x^2(w'/w)^2 + 3x(w'/w) + x^3 - 1}{[2(w'/w)^2 + x][4(w'/w)^3 + 2x(w'/w) - 1]}, \tag{7.45}$$

where w is any solution of Eq. (7.42) and w' its derivative.

7.5.2.2 *An integral equation*

A noteworthy result occurred more recently, with the discovery of an exact linearization of P_{II} by an integral equation [Ablowitz & Segur (1977)]. This result comes from the research by these authors into a connection between the partial derivative equations, which are solved by the inverse scattering transform and the Painlevé equations [Ablowitz & Segur (1981)].

Let us consider the linear integral equation

$$K(x,y) = r\, Ai\left(\frac{x+y}{2}\right) + \sigma\frac{r^2}{4}\int_x^\infty\int_x^\infty K(x,z) Ai\left(\frac{z+s}{2}\right) Ai\left(\frac{s+y}{2}\right) dz ds, \tag{7.46}$$

where $y \geq x$, $\sigma = \pm 1$, and r is a parameter. Then the function $W(z;r) = K(z,z)$ is the solution to the Painlevé equation (case $a = 0$ for Eq. (7.40)),

$$W'' = zW + 2\sigma W^3,$$

under the condition that as $z \to \infty$, W behaves like the Airy function:

$$W(z;r) \approx r\, Ai(z).$$

It can also be seen that the kernel of the integral equation can be written differently. Indeed, by using the formula (3.55), we obtain [Ablowitz & Clarkson (1991)] the Airy kernel

$$\int_x^\infty Ai\left(\frac{z+s}{2}\right) Ai\left(\frac{s+y}{2}\right) \mathrm{d}s = \frac{4}{z-y}\left[Ai\left(\frac{x+z}{2}\right) Ai'\left(\frac{x+y}{2}\right) - Ai'\left(\frac{x+z}{2}\right) Ai\left(\frac{x+y}{2}\right)\right].$$

In relation to the Painlevé equation, this kernel was studied in depth by Tracy & Widom (1994).

7.5.2.3 *Rational solutions*

Kajiwara and Ohta (1996) determined a determinantal representation of rational solutions of $P_{\rm II}$ [Clarkson (2003)].

Theorem 7.1. *Let $p_k(z)$ be the polynomial defined by*

$$\sum_{k=0}^{\infty} p_k(z)\,\lambda^k = \exp\left(z\lambda - \frac{4}{3}\lambda^3\right), \tag{7.47}$$

with $p_k(z) = 0$ for $k < 0$, and $\tau_n(z)$ be the $n \times n$ determinant $(n \geq 1)$

$$\tau_n(z) = \begin{vmatrix} p_n(z) & p_{n+1}(z) & \cdots & p_{2n-1}(z) \\ p_{n-2}(z) & p_{n-1}(z) & \cdots & p_{2n-3}(z) \\ \vdots & \vdots & \ddots & \vdots \\ p_{-n+2}(z) & p_{-n+3}(z) & \cdots & p_1(z) \end{vmatrix}. \tag{7.48}$$

Then

$$w(z;\, n) = \frac{\mathrm{d}}{\mathrm{d}z}\left\{\ln\left[\frac{\tau_{n-1}(z)}{\tau_n(z)}\right]\right\}, \quad n \geq 1$$

satisfies $P_{\rm II}$ (Eq. (7.40)) with $a = n$.

On this important result, we can clearly see that Airy functions are again involved, since the polynomials $p_n(z)$ are (except for a scaling) the Airy polynomials (cf. §4.2.3 Eq. (4.55)).

Exercises

(1) Prove the relation $L = 1.986(IE/q)^{1/3}$ given on page 137. *Hint*: The first zero of the Bessel function $J_{-1/3}(u)$ is $u = 1.866$.

(2) What is the solution to the heat equation with the initial condition $\psi(x) = Ai'(x)$?

(3) Prove that $\int_{-\infty}^{\infty} \mathcal{A}(s)\,\mathrm{d}s = 1$, where $\mathcal{A}(s)$ is defined by Eq. (7.31).

(4) Prove the integral representation of the probability density (7.32). *Hint*: See integrals (3.104) and (3.106).

(5) Check that the following function is a similarity solution to the l-KdV equation

$$\frac{1}{(12t)^{1/6}}\,Ai^2\left[\frac{x}{(12t)^{1/3}}\right].$$

Hint: See Eq. (3.147).

(6) Explore the problem of convective–diffusive mass transfer with chemical reaction in a Couette flow. In particular, at steady state, neglecting diffusion in the direction of convective transport, the problem is governed by the following partial differential equation

$$(u_0 + ay)\frac{\partial c}{\partial x} = D\frac{\partial^2 c}{\partial y^2} - kc,$$

with the boundary conditions

$$c(0, y) = 0, \quad y > 0;$$

$$c(x, 0) = c_0, \quad x \leq 0;$$

$$\lim_{y\to\infty, x>0} c(x, y) < \infty.$$

Hint: See the following articles [Apelblat (1980, 1982); Chen *et al.* (1996); Chen & Arce (1997)].

Chapter 8

Applications to Quantum Physics

8.1 The Schrödinger equation

8.1.1 *Particle in a uniform field*

8.1.1.1 *The stationary case*

Let us consider a free particle of charge q moving on the $\vec{x}$ axis plunged into a uniform electric field $\overrightarrow{\mathcal{E}}$. This particle is submitted to the force $\overrightarrow{F} = q\,\overrightarrow{\mathcal{E}}$ and its potential energy is $U = -Fx$. So the Schrödinger equation is verified by the wave function of the particle

$$\frac{\mathrm{d}^2\psi}{\mathrm{d}x^2} + \frac{2m}{\hbar^2}\left(E + Fx\right)\psi = 0, \tag{8.1}$$

where E is the total energy of the particle. By performing the change of variable

$$\xi = \left(x + \frac{E}{F}\right)\left(\frac{2mF}{\hbar^2}\right)^{1/3},$$

where ξ is a one-dimensional variable, the Schrödinger equation is reduced to the Airy equation (2.1)

$$\frac{\mathrm{d}^2\psi}{\mathrm{d}\xi^2} + \xi\psi = 0. \tag{8.2}$$

The solution to this equation is

$$\psi\left(\xi\right) = N\,Ai\left(-\xi\right) + N'\,Bi\left(-\xi\right),$$

where Ai and Bi are the homogeneous Airy functions. But $Bi(x)$ goes to infinity for $x > 0$. This solution is not relevant, so N' is chosen as 0, reducing the solution to Eq. (8.2) to

$$\psi\left(\xi\right) = N\,Ai\left(-\xi\right).$$

The constant N is determined by the energy normalisation condition for the wave functions of the continuum spectrum [Landau & Lifchitz (1966)]

$$\int_{-\infty}^{+\infty} \psi(\xi)\,\psi^*(\xi')\,\mathrm{d}x = \delta(E - E').$$

Then we obtain

$$N = \frac{(2m)^{1/3}}{\hbar^{2/3}F^{1/6}},$$

and the exact solution of Eq. (8.1) is

$$\psi(x) = \frac{(2m)^{1/3}}{\hbar^{2/3}F^{1/6}} Ai\left[-\frac{(2mF)^{1/3}}{\hbar^{2/3}}\left(x + \frac{E}{F}\right)\right]. \tag{8.3}$$

Of course, this result can be obtained by seeking the wave function $\phi(p)$ in the momentum representation [Landau & Lifchitz (1966)]. With the previous notations, the Hamiltonian operator in the momentum representation is

$$\hat{H} = \frac{p^2}{2m} - \mathrm{i}\hbar F\frac{\mathrm{d}}{\mathrm{d}p}.$$

Then the Schrödinger equation for the wave function $\phi(p)$ is

$$-\mathrm{i}\hbar F\frac{\mathrm{d}\phi(p)}{\mathrm{d}p} + \left(\frac{p^2}{2m} - E\right)\phi(p) = 0.$$

Solving this equation, we get

$$\phi(p) = \frac{1}{\sqrt{2\pi\hbar F}}\mathrm{e}^{\frac{\mathrm{i}}{\hbar F}\left(Ep - \frac{p^3}{6m}\right)},$$

which is energy normalised by the condition

$$\int_{-\infty}^{+\infty} \phi^*_E(p)\phi_{E'}(p)\mathrm{d}p = \delta(E - E').$$

The wave function in the position space is the Fourier transform of the wave function in the momentum space. We thus obtain the solution to (8.3)

$$\psi(x) = \int_{-\infty}^{+\infty} \phi(p)\mathrm{e}^{\frac{\mathrm{i}}{\hbar}px}\mathrm{d}x,$$

where $p = u\,(2m\hbar F)^{1/3}$.

The case of a particle in the potential

$$V(r) = \begin{cases} r & \text{if } r > 0 \\ \infty & \text{if } r \leq 0, \end{cases}$$

is similar to the previous one. However, in this potential well, the states are bounded and the energy spectrum is discrete: the energy E_n corresponds to the level n. The Schrödinger equation of the nth state of the particle wave function is

$$\frac{\mathrm{d}^2\psi_n(r)}{\mathrm{d}r^2} + \frac{2m}{\hbar^2}\left(E_n - r\right)\psi_n(r) = 0. \tag{8.4}$$

Performing the change of variable

$$\xi = \left(E_n - r\right)\left(\frac{2m}{\hbar^2}\right)^{1/3},$$

yields the Airy equation (cf. Eq. (8.2))

$$\frac{\mathrm{d}^2\psi_n}{\mathrm{d}\xi^2} + \xi\psi_n = 0. \tag{8.5}$$

The solution is (excluding the Bi term)

$$\psi_n\left(\xi\right) = N\,Ai\left(-\xi\right).$$

The energy levels are determined by the condition $\psi_n(0) = 0$

$$Ai\left[-\left(\frac{2m}{\hbar^2}\right)^{1/3} E_n\right] = 0.$$

Then the E_n are determined by the zeros a_n of the Airy function (cf. §2.2.1)[1]

$$E_n = -a_{n+1}\left(\hbar^2/2m\right)^{1/3},$$

i.e.

$$\begin{aligned} E_0 &\simeq 2.33811\left(\hbar^2/2m\right)^{1/3} \\ E_1 &\simeq 4.08795\left(\hbar^2/2m\right)^{1/3} \\ E_2 &\simeq 5.52056\left(\hbar^2/2m\right)^{1/3} \\ E_3 &\simeq 6.78671\left(\hbar^2/2m\right)^{1/3} \\ &\vdots \end{aligned}$$

[1]We note: $E_n \propto -a_{n+1}$, and not: $E_n \propto -a_n$, because the numbering of the levels begins from zero for the ground state, whereas a_n is the $\mathrm{n^{th}}$ zero of Ai.

The normalisation factor N is determined by the orthonormalisation condition of the wave functions

$$\int_0^{+\infty} \psi_n(r)\psi_m^*(r)\mathrm{d}r = \delta\left(E_n - E_m\right),$$

with the wave function ψ_n written

$$\psi_n(r) = N\, Ai\left[r\left(\frac{2m}{\hbar^2}\right)^{1/3} + a_n\right].$$

We then obtain (cf. §4.4) $N = \left(\frac{2m}{\hbar^2}\right)^{1/6}\frac{1}{Ai'(a_n)}$. The solution to Eq. (8.4) is finally given by

$$\psi_n(r) = \left(\frac{2m}{\hbar^2}\right)^{1/6}\frac{1}{Ai'\left(a_n\right)} Ai\left[\left(\frac{2m}{\hbar^2}\right)^{1/3}\left(r - E_n\right)\right]. \tag{8.6}$$

We can see on Fig. (8.1) the potential $V(r) = r$ and the first energy levels, with $m = 1/2$ and $\hbar = 1$.

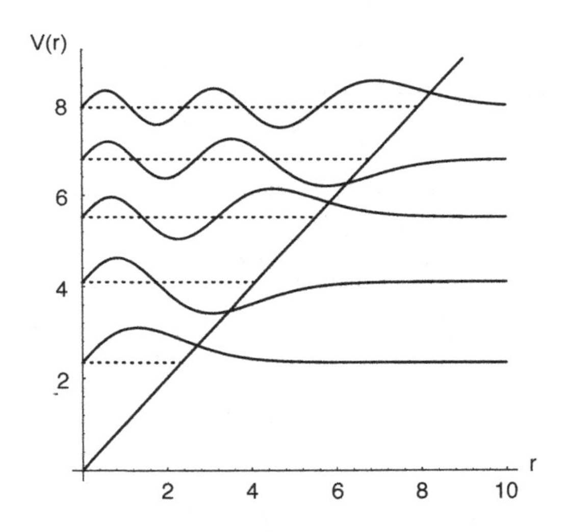

Fig. 8.1 The linear potential: this figure shows the first energy levels and the corresponding wave functions.

8.1.1.2 *The time dependent case*

In this case, we have to solve the time dependent Schrödinger equation

$$\mathrm{i}\hbar\frac{\partial\Psi}{\partial t} = -\frac{\hbar^2}{2m}\frac{\partial^2\Psi}{\partial x^2} + x\Psi. \tag{8.7}$$

We first separate the variables $\Psi(x,t) = \mathrm{e}^{-\mathrm{i}Et/\hbar}\,\psi(x)$ leading to the preceding equation for $\psi(x)$ (Eq. (8.5)). Then the general solution of the time dependent Schrödinger equation is given by

$$\Psi(x,t) = \sum_{n=0}^{\infty} c_n \mathrm{e}^{-\mathrm{i}E_n t/\hbar}\,\psi_n(x), \tag{8.8}$$

where $\psi_n(x)$ is the Airy function

$$\psi_n(x) = \left(\frac{2m}{\hbar^2}\right)^{1/6} \frac{1}{Ai'(a_n)} Ai\left[\left(\frac{2m}{\hbar^2}\right)^{1/3}(x - E_n)\right],$$

E_n being related to the zero of the Airy function a_n. Therefore, the general solution $\Psi(x,t)$ is expressed (except for some extra factors) as a Fourier–Airy series (see §4.4), while the coefficients c_n are calculated with the initial condition $\Psi(x,0)$ from the integral

$$c_n = \int_0^{\infty} \Psi(x,0)\psi_n(x)\mathrm{d}x.$$

8.1.2 *The $|x|$ potential*

Let us consider the one-dimensional system with a particle in the potential $V(x) = |x|$, $x \in \mathbb{R}$. The bound states are determined by solving the Schrödinger equation for the wave function $\psi_n(x)$ of the particle in the state n:

$$\frac{\mathrm{d}^2\psi_n(x)}{\mathrm{d}x^2} + \frac{2m}{\hbar^2}\left(E_n - |x|\right)\psi_n(x) = 0. \tag{8.9}$$

Proceeding in the same way as above, for the particle in the potential $V(r) = r$, we have, for $x > 0$, the wave function

$$\psi_n(x) = N\,Ai\left[\left(\frac{2m}{\hbar^2}\right)^{1/3}(x - E_n)\right], \tag{8.10}$$

where the normalisation constant N and the energy levels E_n remain to be determined.

The energy levels are defined by matching two wave functions $\psi_n^{(1)}(x)$ and $\psi_n^{(2)}(x)$ at $x = 0$. At this step, the respective wave functions and their derivatives must be equal (except for the sign)

$$\psi_n^{(1)}(0) = \psi_n^{(2)}(0) \tag{8.11}$$
$$\psi_n^{'(1)}(0) = \pm\psi_n^{'(2)}(0). \tag{8.12}$$

It is then necessary to distinguish two cases, according to the parity of the quantum number n:

▶ n is even: the wave function has an even number of nodes, the axis $x = 0$ is a symmetry axis, and consequently the derivative of $\psi_n(x)$ is zero for $x = 0$ (i.e. $\psi_n(x)$ presents a local extremum in $x = 0$, cf. Fig. (8.2)). We thus have (with the atomic units $2m = \hbar = 1$ to simplify) $\psi_n^{(1)}(x) = N\,Ai\,(-x - E_n)$, $x < 0$, and $\psi_n^{(2)}(x) = N\,Ai\,(x - E_n)$, $x > 0$. The relation (8.12) enables us to obtain

$$Ai'\left[-E_n\right] = 0,$$

i.e.

$$E_n = -a'_{n+1},$$

where a'_n indicates the nth zero of the Ai' function (cf. §2.2.1).

▶ n is odd: the wave function has an odd number of nodes, which are located on the $x = 0$ axis. There is no longer an axial symmetry, but a central symmetry at the origin (cf. Fig. (8.2)). The wave functions are $\psi_n^{(1)}(x) = -N\,Ai\,(-x - E_n)$, $x < 0$, and $\psi_n^{(2)}(x) = N\,Ai\,(x - E_n)$, $x > 0$. Equation (8.11) gives us

$$Ai\left[-e_n\right] = 0,$$

i.e.

$$E_n = -a_{n+1},$$

where a_n is the nth zero of the Ai function, already encountered in the case of the potential $V(r) = r$.

The coefficient of normalisation N is determined by the condition

$$\int_{-\infty}^{+\infty} \psi_n(r)\psi_m^*(r)dr = \delta\left(E_n - E_m\right),$$

which is reduced to

$$N^2 \int_0^{+\infty} Ai\,(x - E_n)\,Ai\,(x - E_m)\,\mathrm{d}x = \delta\left(E_n - E_m\right).$$

Then we obtain (cf. §4.4 and formula (3.51))

$$N = \frac{1}{\sqrt{-a'_n}Ai(a'_n)},$$

if n is even, and

$$N = \frac{1}{Ai'(a_n)},$$

if n is odd.

Finally, we reach the solution of the Schrödinger Eq. (8.9) by restoring the constant $(2m/\hbar^2)^{1/3}$.

► n even:

$$\psi_n(x) = \left(\frac{2m}{\hbar^2}\right)^{1/6} \frac{1}{\sqrt{-a'_n}Ai(a'_n)} Ai\left[\left(\frac{2m}{\hbar^2}\right)^{1/3}(|x| - E_n)\right], \tag{8.13}$$

that is to say

$$E_n = -a'_{n+1}\left(\frac{\hbar^2}{2m}\right)^{1/3}.$$

► n odd:

$$\psi_n(x) = \text{sgn}(x)\left(\frac{2m}{\hbar^2}\right)^{1/6} \frac{1}{Ai'(a_n)} Ai\left[\left(\frac{2m}{\hbar^2}\right)^{1/3}(|x| - E_n)\right], \tag{8.14}$$

that is to say

$$E_n = -a_{n+1}\left(\frac{\hbar^2}{2m}\right)^{1/3}.$$

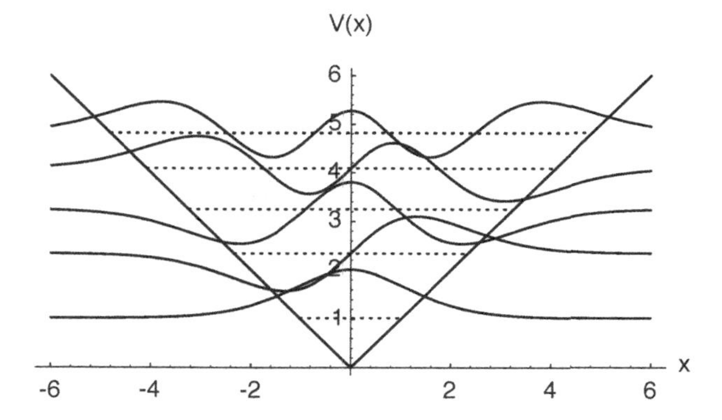

Fig. 8.2 The $|x|$ potential. On this figure, the first energy levels and the corresponding wave functions are represented.

Figure (8.2) shows the potential $V(x) = |x|$ and the first energy levels, with $m = 1/2$ and $\hbar = 1$.

8.1.3 *Uniform approximation of the Schrödinger equation*

The one-dimensional Schrödinger equation (or the radial equation in the three-dimensional case) can be written [Berry & Mount (1972); Eu (1984)]

$$y'' + \frac{p^2(r)}{\hbar^2} y = 0, \quad r \in [0, +\infty), \tag{8.15}$$

where $p(r)$ is the momentum of the particle. We find here Eq. (5.5) where $q_1(x) = p^2(r)$, $q_2(x) = 0$ and $\lambda = 1/\hbar$. Proceeding as in §5.2, we carry out the changes of variable and function according to

$$\begin{cases} x = x(r) \\ y = z\Omega(x). \end{cases}$$

Equation (8.15) then becomes

$$\begin{aligned} z'' &+ \left[2\frac{\mathrm{d}}{\mathrm{d}x}\ln\Omega - \frac{\mathrm{d}^2 r}{\mathrm{d}x^2}\frac{\mathrm{d}x}{\mathrm{d}r}\right] z' \\ &+ \left[\frac{1}{\Omega}\frac{\mathrm{d}^2\Omega}{\mathrm{d}x^2} - \frac{\mathrm{d}^2 r}{\mathrm{d}x^2}\frac{\mathrm{d}x}{\mathrm{d}r}\left(\frac{\mathrm{d}}{\mathrm{d}x}\ln\Omega\right) + \frac{p^2(r)}{\hbar^2}\left(\frac{\mathrm{d}x}{\mathrm{d}r}\right)^2\right] z = 0, \end{aligned} \tag{8.16}$$

where $z' = \mathrm{d}z/\mathrm{d}x$ and $z'' = \mathrm{d}^2 z/\mathrm{d}x^2$. We choose the change of function so that the term z' disappears. We then obtain

$$\Omega = \left(\frac{\mathrm{d}x}{\mathrm{d}r}\right)^{-1/2}.$$

Thus we are able to transform Eq. (8.16) into an equation depending only on x and p where the Schwarzian derivative appears, $\{x, r\}$:

$$\{x, r\} = \frac{x'''}{x'} - \frac{3}{2}\left(\frac{x''}{x'}\right)^2,$$

in which x', x'', x''' are the successive derivatives of x as a function of r. We thus obtain

$$z'' + \left[\frac{p^2(r)}{\hbar^2 x'^2(r)} - \frac{1}{x'^2(r)}\{x, r\}\right] z = 0. \tag{8.17}$$

In the semiclassical limit (the Planck constant $\hbar$ tends formally towards 0), the term containing the Schwarzian derivative can be neglected compared to the other term. The generalised JWKB approximation consists of preserving only the first term in z, with a suitable choice for change of variable

$$z'' + \frac{p^2(r)}{\hbar^2 x'^2} z = 0. \tag{8.18}$$

This choice will be driven by considerations under the topological nature (in particular the turning points) of the momentum $p(r)$. The general method consists in choosing

$$\xi(x) = \frac{p^2(r)}{\hbar^2 x'^2}, \tag{8.19}$$

so that we know the solutions of Eq. (8.18). Then the integration of Eq. (8.19) determines the change of variable

$$\int_{x_t}^{x} \xi^{1/2}(x)\mathrm{d}x = \frac{1}{\hbar}\int_{r_t}^{r} p(u)\mathrm{d}u, \tag{8.20}$$

where r_t is a zero of the momentum $p(r)$, or turning point. So, according to the change of function (8.1.3), the approximate solution of the Schrödinger equation is now

$$y = \left(\frac{\mathrm{d}x}{\mathrm{d}r}\right)^{-1/2} z(x). \tag{8.21}$$

8.1.3.1 *The JWKB approximation*

This approximation consists of considering Eq. (8.18) as a harmonic oscillator and consequently setting $\xi(x) = 1$. The integration of Eq. (8.19) gives us

$$x(r) = \frac{1}{\hbar}\int_{r_t}^{r} p(u)\mathrm{d}u,$$

if $p^2(r) > 0$. The wave function, which is the approximate solution of the Schrödinger equation, is thus written

$$y(r) = \frac{C_1}{\sqrt{p(r)}}\mathrm{e}^{-\frac{1}{\hbar}\int_{r_t}^{r} p(u)\mathrm{d}u} + \frac{C_2}{\sqrt{p(r)}}\mathrm{e}^{\frac{1}{\hbar}\int_{r_t}^{r} p(u)\mathrm{d}u}.$$

The constants C_1 and C_2 are determined by the conditions of normalisation and by the connection of the exact solutions to the Schrödinger equation [Landau & Lifchitz (1966)] in the semiclassical limit. The latter condition gives

$$y(r) = \frac{C}{\sqrt{p(r)}}\sin\left(\frac{1}{\hbar}\int_{r_t}^{r} p(u)\mathrm{d}u + \frac{\pi}{4}\right) \tag{8.22}$$

in the classically allowed region.

In the classically forbidden region ($p^2(r) < 0$), we deduce from Eq. (5.15)

$$y(r) = \frac{1}{\sqrt{-p(r)}} \left\{ C_3 \mathrm{e}^{\frac{1}{\hbar}\int_{r_t}^{r} p(u)\mathrm{d}u} + C_4 \mathrm{e}^{-\frac{1}{\hbar}\int_{r_t}^{r} p(u)\mathrm{d}u} \right\}.$$

In this equation, we cannot physically keep the term going towards infinity as $r \to \infty$. Then the JWKB solution to Eq. (8.15) is

$$y(r) = \frac{C'}{\sqrt{-p(r)}} \mathrm{e}^{-\frac{1}{\hbar}\int_{r_t}^{r} p(u)\mathrm{d}u} . \tag{8.23}$$

Physically, the criterion of validity of the JWKB approximation can be expressed in two forms

i) $\frac{1}{2\pi}\left|\frac{\mathrm{d}\lambda}{\mathrm{d}r}\right| \ll 1$, i.e. λ, the de Broglie wavelength of the particle, must vary slowly at a distance about this same wavelength;
ii) $\frac{\hbar m|F|}{p^3} \ll 1$, where $F = -\mathrm{d}V/\mathrm{d}x$ is the classical force acting on the particle. This shows, in particular, that the JWKB approximation is no longer valid at the turning points, i.e. at the points where the momentum $p(r)$ is null.

8.1.3.2 *The Airy uniform approximation*

In this section we present an approximation to the solutions of Eq. (8.15) when the momentum $p(r)$ has a single turning point r_t. For this purpose we now make the change of variable $\xi(x) = x$. Under this condition, the solutions of the differential equation (8.18) are Airy functions. Taking into account the behaviour at infinity (cf. Figs. (2.2) and (2.3)), the $Ai(x)$ function is the only admissible function. The integration of Eq. (8.19) then gives, for the change of variable,

$$x(r) = \left[\frac{3}{2\hbar} \int_{r_t}^{r} p(u)\mathrm{d}u \right]^{2/3} . \tag{8.24}$$

Therefore, except for a multiplicative constant, the approximate solution of the Schrödinger equation, uniformly valid far from or close to the turning point, may be written

$$y(r) = \left[\hbar^2 \frac{x(r)}{p^2(r)} \right]^{1/4} Ai\left[\varepsilon x(r)\right], \tag{8.25}$$

where ε is -1 in the classically allowed region and $+1$ in the classically forbidden region. If there are several turning points, a good connection between wave functions such as (8.25) can accurately approach the exact solution of the Schrödinger equation [Miller (1968)]. Note moreover that, by taking the asymptotic limit of the expression (8.25), with the formulae (2.45) and (2.50), we find the usual JWKB semiclassical limit given by the formulae (8.22) and (8.23).

Note also that $x(r)$ verifies the relation:

$$\frac{p(r)}{\hbar\sqrt{x(r)}} = x'(r).$$

Consequently, the expression (8.25) of the uniform solution to the Schrödinger equation can be written

$$y(r) = \frac{1}{\sqrt{x'(r)}} Ai\left[x(r)\right]. \tag{8.26}$$

In addition, the expansion of $x'(r)$ in the neighbourhood of the turning point r_t

$$x'^2(r) = \frac{p^2(r)}{\hbar^2 x(r)} \approx \frac{0 + (r - r_t)\left(p^2(r_t)\right)'}{0 + \hbar^2 (r - r_t)\, x'(r_t)} = \frac{2mV'(r_t)}{\hbar^2 x'(r_t)},$$

leads for the wave function, to the following important expression

$$y(r) = \frac{1}{\sqrt{\alpha}} Ai\left[\alpha\,(r - r_t)\right], \tag{8.27}$$

for $r \approx r_t$, where α is a constant depending on the slope of the potential at the turning point

$$\alpha = \left[2mV'(r_t)\right]^{1/3} \hbar^{-2/3},$$

m being the reduced mass of the system.

8.1.3.3 *Exact vs approximate wave functions*

To highlight the validity of the Airy uniform approximation, we shall consider two well-known potentials for which we can solve analytically the Schrödinger equation. We shall be able, therefore, to compare the exact wave functions and the uniform ones.

► The case of the exponential potential:

Consider a wave function $\psi_E(r)$, satisfying the Schrödinger equation

$$\frac{\mathrm{d}^2\psi_E(r)}{\mathrm{d}r^2} + \frac{p^2(r)}{\hbar^2}\psi_E(r) = 0, \tag{8.28}$$

where the momentum $p(r)$ is defined by

$$p^2(r) = 2m\left[E - V(r)\right].$$

If we choose a purely repulsive exponential potential

$$V(r) = V_0\ \mathrm{e}^{-r/r_m},$$

the exact solution of the Schrödinger equation is [Child (1974)]

$$\psi_{_E}(r) = \frac{2}{\pi}\left[\frac{mr_m}{\hbar^2}\sinh\left(2\pi a\right)\right]^{1/2} K_{\,\mathrm{i}2a}\left(2b\mathrm{e}^{-r/2r_m}\right), \tag{8.29}$$

where $K_\nu(z)$ is the modified Bessel function, and

$$\begin{cases} a = \frac{r_m}{\hbar}\sqrt{2mE} \\ b = \frac{r_m}{\hbar}\sqrt{2mV_0}\,. \end{cases}$$

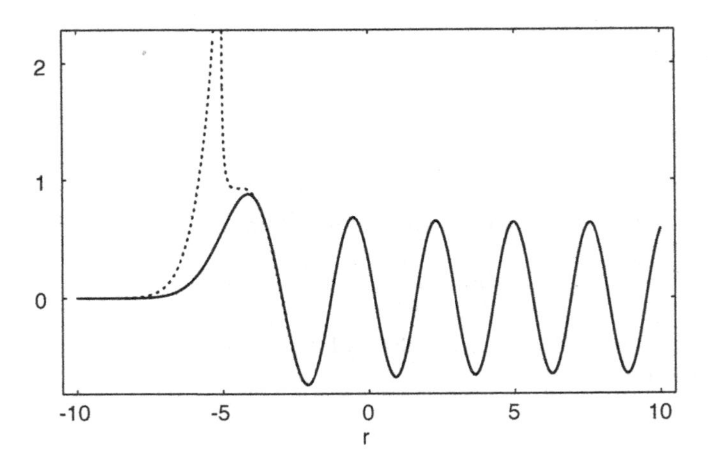

Fig. 8.3 Exponential potential: comparison between the exact wave function (solid line) and the JWKB wave function (dotted line).

As this system admits a single turning point, we can apply the uniform approximation method to Eq. (8.28). The solution is therefore

$$\psi_{_E}^{\mathrm{unif}}(r) = \sqrt{2m}\left[\frac{\hbar^2 x(r)}{p^2(r)}\right]^{1/4} Ai\left[\varepsilon x(r)\right], \tag{8.30}$$

where $\varepsilon = +1$ in the classically forbidden regions and $\varepsilon = -1$ in the classically allowed regions, $x(r)$ being defined by the relation (8.24).

On Figs. (8.3) to (8.4), we compare the exact wave function, formula (8.29) (solid line), with the JWKB wave function, formulae (8.22) and

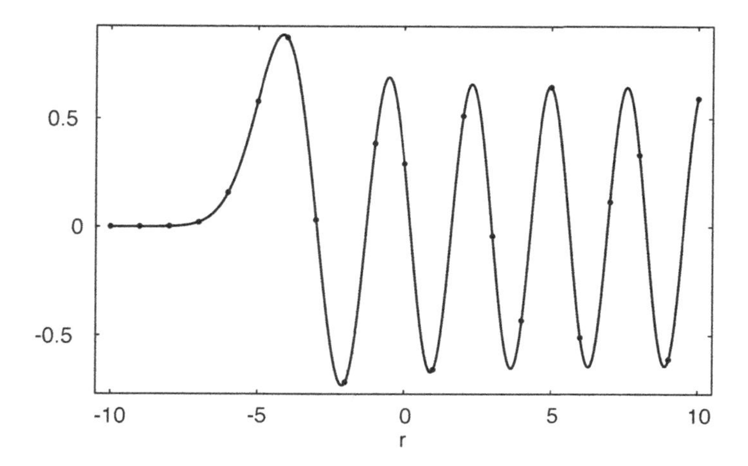

Fig. 8.4 Exponential potential: comparison between exact wave function (solid line) and uniform wave function (•).

(8.23) (dotted lines), and with the uniform wave function, formula (8.30) (bullets •).

On Fig. (8.3), it can be seen that the JWKB wave function becomes divergent at the turning point. On Fig. (8.4), the Airy wave function is continuous, even in the neighbourhood of this turning point, the agreement with the exact function being of the order of one percent.

► The case of the Coulomb potential:

The Schrödinger radial equation associated with an electron in the Coulomb potential $V(r) = 1/r$ of a hydrogen atom is, for the discrete spectrum ($E = -1/n^2$) (in atomic units) [Landau & Lifchitz (1966)],

$$\frac{\mathrm{d}^2\psi_{n,\ell}(r)}{\mathrm{d}r^2} + \frac{2}{r}\frac{\mathrm{d}\psi_{n,\ell}(r)}{\mathrm{d}r} + \left(\frac{2}{r} - \frac{(\ell+1/2)^2}{r^2} - \frac{1}{n^2}\right)\psi_{n,\ell}(r) = 0. \quad (8.31)$$

The exact solution of this equation is

$$\psi_{n,\ell}(r) = -\frac{2}{n^2}\sqrt{\frac{(n-\ell-1)!}{[(n+\ell)!]^3}}\ \mathrm{e}^{-r/n}\left(\frac{2r}{n}\right)^{\ell} L_{n+\ell}^{2\ell+1}\left(\frac{2r}{n}\right), \quad (8.32)$$

where $L_\beta^\alpha(x)$ is the generalised Laguerre polynomials [Abramowitz & Stegun (1965)]. The classical turning points r_+ and r_- are defined by

$$r_\pm = n^2 \pm n\sqrt{n^2-(\ell+1/2)^2}.$$

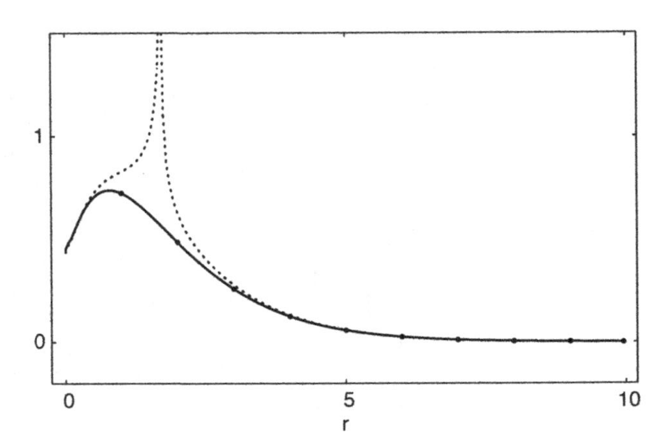

Fig. 8.5 Coulomb potential: comparison between the exact (solid line), JWKB (dotted line) and uniform wave functions (•) for $n = 1$, $\ell = 0$.

Applying the uniform approximation to Eq. (8.31), we obtain the solution [Fonck & Tracy (1980)]

$$\psi_{n,\ell}^{\text{unif}}(r) = N_n \left(\frac{6\pi}{p_s(r)}\right)^{1/2} \left(\frac{2}{3}\right)^{1/3} [x_s(r)]^{1/6} Ai\left\{(-1)^s \left[\frac{3}{2}x_s(r)\right]^{2/3}\right\}, \tag{8.33}$$

with $r > r_-$, and

- $s = 1$ for $r_- < r \leq r_+$ and $s = 2$ for $r \geq r_+$;
- $p_1^2(r) = -p_2^2(r) = \frac{2}{r} - \frac{(\ell+1/2)^2}{r^2} - \frac{1}{n^2}$;
- $$\begin{aligned} x_1(r) &= \int_r^{r_+} p_1(r)\mathrm{d}r \\ &= -rp_1(r) - n\arcsin\left(\frac{r-n^2}{n\sqrt{n^2-(\ell+1/2)^2}}\right) \\ &\quad + \left(\ell+\tfrac{1}{2}\right)\arcsin\left(\frac{n}{r}\frac{r-(\ell+1/2)^2}{\sqrt{n^2-(\ell+1/2)^2}}\right) + \left(n-\ell-\tfrac{1}{2}\right)\frac{\pi}{2}; \end{aligned}$$
- $$\begin{aligned} x_2(r) &= \int_{r_+}^{r} p_2(r)\mathrm{d}r \\ &= rp_2(r) - n\ln\left|rp_2(r) + \tfrac{r}{n} - n\right| \\ &\quad + \tfrac{n}{2}\ln\left|n^2 - \left(\ell+\tfrac{1}{2}\right)^2\right| - \left(\ell+\tfrac{1}{2}\right)\ln\left|\frac{rp_2(r)+(\ell+1/2)}{r} - \frac{1}{(\ell+1/2)}\right| \\ &\quad + \tfrac{1}{2}\left(\ell+\tfrac{1}{2}\right)\ln\left|\frac{1}{(\ell+1/2)^2} - \frac{1}{n^2}\right|; \end{aligned}$$
- $N_n = (-1)^{n-\ell-1}\left(2\pi n^3\right)^{-1/2}$, is the normalisation coefficient given by Bethe & Salpeter (1957).

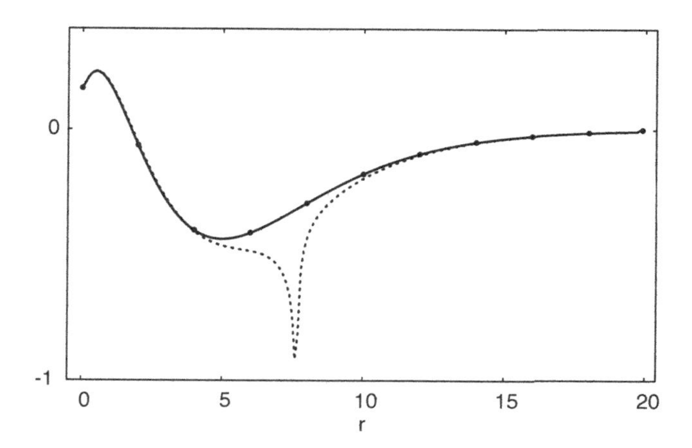

Fig. 8.6 Coulomb potential: comparison between the exact (solid line), JWKB (dotted line) and uniform wave functions (•) for $n = 2$, $\ell = 0$.

Moreover, from Eqs. (8.22) and (8.23), we can deduce the JWKB solutions to Eq. (8.31)

$$\begin{cases} \psi_{n,\ell}(r) = N_n \frac{2}{\sqrt{p_1(r)}} \cos\left[x_1(r) - \frac{\pi}{4}\right] & \text{for} \quad r_- < r \leq r_+ \\ \psi_{n,\ell}(r) = N_n \frac{1}{\sqrt{p_2(r)}} \mathrm{e}^{-x_2(r)} & \text{for} \quad r > r_+ \, , \end{cases}$$

where p_1, p_2, x_1 and x_2 were previously defined.

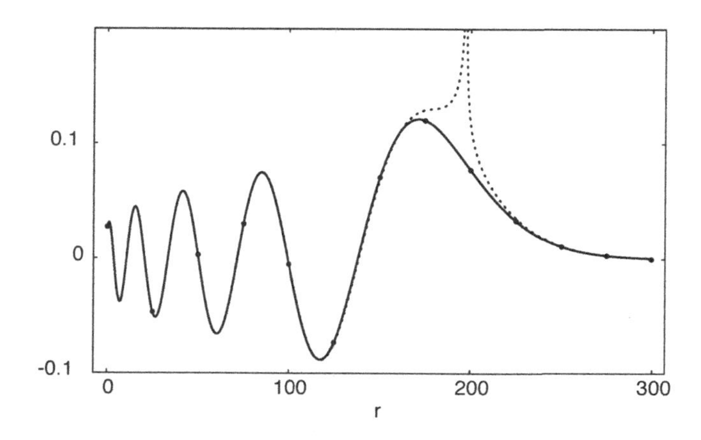

Fig. 8.7 Coulomb potential: comparison between the exact (solid line), JWKB (dotted line) and uniform wave functions (•) for $n = 10$, $\ell = 1$.

On Figs. (8.5) to (8.7), we compare the exact wave function, formula (8.32) (solid line), the JWKB wave function, formula (8.1.3.3) (dotted lines), and the uniform wave function, formula (8.33) (bullets •). As in the case of the exponential potential, on the first two figures it can be seen that the JWKB function becomes divergent at the turning point r_+, whereas the uniform wave function is continuous.

8.2 Evaluation of the Franck–Condon factors

In this section, we shall establish the JWKB semiclassical expression and the uniform semiclassical expression, of

$$F_{if} = \langle \psi_i(r) \left| D(r) \right| \psi_f(r) \rangle = \int \psi_i(r) D(r) \psi_f(r) \mathrm{d}r, \tag{8.34}$$

where $\psi_i(r)$ is the wave function describing an initial quantum state, and $\psi_f(r)$ a final state, with $D(r)$ being the dipole operator.

F_{if} is the Franck–Condon factor, named after the work of J. Franck and E. U. Condon (Condon (1928); see also Child (1974) and Tellinghuisen (1985)). This factor is used, in particular, to calculate line profiles or transition probabilities between two levels: i (initial) and f (final). For example, within the framework of an adiabatic theory, the line profile corresponding to the transition between a level i (state described by the wave function ψ_i) and a level f (state described by the wave function ψ_f) is given by the average of the square of the dipole matrix element of the considered transition. This is nothing else but the application of the Fermi golden rule, i.e.

$$J(\omega) = \int_0^\infty \mathrm{d}p_i \; \rho(p_i) \sum_\ell (2\ell + 1) \left| \langle \psi_i(r) \left| D(r) \right| \psi_f(r) \rangle \right|^2, \tag{8.35}$$

where the wave functions verify the Schrödinger radial equations ($\alpha = i, f$)

$$\frac{\mathrm{d}^2 \psi_\alpha(r)}{\mathrm{d}r^2} + p_\alpha^2(r) \psi_\alpha(r) = 0,$$

with $p_\alpha^2(r) = 2m \left(\varepsilon_\alpha - V_\alpha(r) - \frac{\ell(\ell+1)}{2mr^2} \right)^{1/2}$ and where ℓ is the angular quantum number, $\rho(p_i)$ being the statistical distribution of the initial states.

Note that F_{if} is *stricto sensu* the Franck–Condon amplitude, whereas the Franck–Condon factor is the square of F_{if}. Both terms are used, however, albeit incorrectly, to indicate F_{if}.

In this section we shall use the atomic units $\hbar = m_e = e = 1$.

8.2.1 *The Franck–Condon principle*

The Franck–Condon principle was established to describe the transition between two electronic states of a molecule. For example, the phenomenon of bound-free transition e.g. the photodissociation of the molecule I_2 in the visible spectrum. This principle stipulates that the relative positions and the momenta of the atoms are preserved during the electronic transition. Physically, this is due to the fact that the electronic velocities are large compared to the nuclear velocities. The Franck–Condon principle also stipulates that it is possible to evaluate the transition probabilities with the help of overlap integrals, containing the classically allowed and forbidden regions. The dipole operator varies slowly with the internuclear distance r, whereas the electronic motion varies very quickly, so that we can replace the dipole operator $D(r)$ by its average value $D(r_0)$, where r_0 is the interatomic distance at the instant of transition. This is realistic since the molecule is massive. The Franck–Condon principle is equivalent to the principle of stationary phase. A Taylor expansion of the dipole operator D is:

$$D(r) = D(r_0) + (r - r_0)\, D'(r_0) + \frac{(r - r_0)^2}{2} D''(r_0) + \dots . \tag{8.36}$$

So we have to the zero order

$$\langle \psi_i(r)\, |D(r)|\, \psi_f(r) \rangle \approx D(r_0)\, \langle \psi_i(r) \mid \psi_f(r) \rangle \,,$$

where r_0 is a stationary phase point.

8.2.2 *The JWKB approximation of Franck–Condon factors*

The calculation of F_{if} requires knowledge of the wave functions of both involved states. If possible, we use the JWKB approximation (see the specific criteria of validity for this method, Landau & Lifchitz (1966)). In this case, in the classically allowed region, we obtain for $\psi_i(r)$ and $\psi_f(r)$ the expressions, (cf. §5.2.1 and §8.1.3.1)

$$\psi_\alpha(r) = \frac{1}{\sqrt{p_\alpha(r)}} \cos\left(\int_{r_\alpha}^{r} p_\alpha(u) \mathrm{d}u - \frac{\pi}{4} \right), \quad \alpha = i, f, \tag{8.37}$$

where r_i and r_f are the turning points of the two considered levels (i.e. $p_\alpha(r_\alpha) = 0$). In order to simplify the notation, the normalisation factor of the wave functions has not been introduced in this formula. For the Franck–Condon factor, we then obtain [Jablonski (1945); Landau & Lifchitz (1966)]

$$F_{if} = \int_{r_m}^{+\infty} \frac{D(r)\mathrm{d}r}{\left(p_i(r)p_f(r)\right)^{1/2}} \times \cos\left(\int_{r_i}^{r} p_i(u)\mathrm{d}u - \frac{\pi}{4}\right) \cos\left(\int_{r_f}^{r} p_f(u)\mathrm{d}u - \frac{\pi}{4}\right),$$

where $r_m = \max(r_i, r_f)$.

This product of *cosine* functions can be changed into a sum of *cosines* where the sum and the difference of the phases appear. Only the term containing the difference of the phases contributes significantly to the integral, the other term oscillating very quickly with r [Jablonski (1945)]. This yields

$$F_{if} \approx \frac{1}{2} \int_{r_m}^{+\infty} \frac{D(r)\mathrm{d}r}{\left(p_i(r)p_f(r)\right)^{1/2}} \times \cos\left(\int_{r_i}^{r} p_i(u)\mathrm{d}u - \int_{r_f}^{r} p_f(u)\mathrm{d}u\right). \tag{8.38}$$

In order to calculate this integral, we shall apply the stationary phase method presented in §5.1. Consider a single stationary phase point r_0, and assume

$$x_\alpha(r) = \int_{r_\alpha}^{r} p_\alpha(u)\mathrm{d}u. \tag{8.39}$$

The second order Taylor series in the neighbourhood of r_0 can be written

$$\begin{aligned} x_i(r) - x_f(r) = {} & x_i(r_0) - x_f(r_0) \\ & + (r - r_0)\left(x_i'(r_0) - x_f'(r_0)\right) \\ & + \frac{(r - r_0)^2}{2}\left(x_i''(r_0) - x_f''(r_0)\right) + \ldots, \end{aligned} \tag{8.40}$$

where $x_i'(r_0) - x_f'(r_0) = p_i(r_0) - p_f(r_0)$. As r_0 is a stationary phase point, the equality of the moments gives

$$p_i(r_0) = p_f(r_0),$$

i.e., using the definition of $p_\alpha(r)$

$$\omega = V_i(r_0) - V_f(r_0) = \Delta V(r_0), \quad (\hbar = 1), \tag{8.41}$$

where $\omega = \varepsilon_i - \varepsilon_f$. Then r_0 is a function depending only on ω. The expression (8.38) can be now written

$$F_{if} \approx \frac{1}{2}\frac{D(r_0)}{p_i(r_0)} \int_{-\infty}^{+\infty} \cos\left(\beta + \gamma\xi^2\right) \mathrm{d}\xi,$$

where $\xi = r - r_0$, and β, γ being functions of r_0:

$$\beta = \frac{mr_0^2}{2p_0} \Delta V'(r_0) \tag{8.42}$$

$$\gamma = \frac{m}{2p_0} \Delta V'(r_0). \tag{8.43}$$

The lower limit of integration can be extended until $-\infty$, since the contribution of the region lower than $r_m - r_0$ is negligible. Moreover, in agreement with the Franck–Condon principle, we replace $D(r)$ by its value at the stationary phase point $D(r_0)$. We finally obtain the expression of the Franck–Condon factor [Landau & Lifchitz (1966)]

$$F_{if} \approx \frac{1}{2}\frac{D(r_0)}{p_i(r_0)} \sqrt{\frac{\pi}{\gamma}} \cos\left(\beta + \frac{\pi}{4}\right). \tag{8.44}$$

The square of the modulus of this factor is written

$$|F_{if}|^2 \approx \frac{1}{8}\frac{D^2(r_0)}{p_i^2(r_0)}\frac{\pi}{\gamma}, \tag{8.45}$$

where we made use of the random phase approximation: the square of the *cosine* is replaced by its average value 1/2. Then we can calculate the profile (8.35) according to the method detailed in the next section (§8.2.3), i.e. by taking as distribution of the initial states $\rho(\varepsilon_i) = \mathrm{e}^{-\varepsilon_i/kT}$ and by changing the discrete summation over the angular quantum numbers into a continuous summation.

This calculation leads to the so-called quasi-static profile [Jablonski (1945)],

$$J(\omega) = \pi D^2(r_0) r_0^2 \, \mathrm{e}^{-V_i(r_0)/kT} \frac{1}{|\Delta V'(r_0)|}, \tag{8.46}$$

except for a multiplicative factor, since the wave functions have not been normalized. Let us recall that ω is related to r_0 by the expression (8.41).

Sando & Wormhoudt continue the expansion (8.40) up to the third order [Sando & Wormhoudt (1973)] and obtain an expression involving the Airy functions for the profile $J(\omega)$. However we shall not detail this calculation, since the uniform approximation will also lead us to a similar expression.

8.2.3 *The uniform approximation of Franck–Condon factors*

We shall now present the complete analytical calculation of the profile using the Airy uniform approximation, starting from the expression of $J(\omega)$,

$$J(\omega) = \int_0^\infty \mathrm{d}\varepsilon_i \, \mathrm{e}^{-\varepsilon_i/kT} \sum_\ell (2\ell+1) \left| \langle \psi_i(r) \left| D(r) \right| \psi_f(r) \rangle \right|^2. \tag{8.47}$$

We consider as wavefunctions (cf. §8.1.3.2):

$$\psi_\alpha(r) = \left(\frac{\pi^2 \sigma_\alpha(r)}{p_\alpha^2(r)} \right)^{1/4} Ai\left[\sigma_\alpha(r)\right], \tag{8.48}$$

where $\alpha = i, f$, and where σ_α is defined by:

$$\sigma_\alpha(r) = \left[\frac{3}{2} \int_{r_\alpha}^{r} p_\alpha(u) \mathrm{d}u \right]^{2/3} = \left[\frac{3}{2} x_\alpha(r) \right]^{2/3}.$$

This approach, which avoids the problems of divergence involved in the JWKB wave functions in the neighbourhood of the turning points, is very close to the method of Bienniek (1977) which also uses the uniform approximation. We have to calculate the matrix element

$$\begin{aligned} F_{if} &= \langle \psi_i(r) \left| D(r) \right| \psi_f(r) \rangle \\ &= \int_{-\infty}^{+\infty} \pi \left(\frac{\sigma_i(r)\sigma_f(r)}{p_i^2(r) p_f^2(r)} \right)^{1/4} D(r) Ai\left[\sigma_i(r)\right] Ai\left[\sigma_f(r)\right] \mathrm{d}r. \end{aligned}$$

We then employ the stationary phase method, since with the definition (2.21) of $Ai(x)$, the preceding integral is written

$$\begin{aligned} F_{if} = \frac{1}{4\pi^2} \iiint_{\mathbb{R}^3} \pi \left(\frac{\sigma_i(r)\sigma_f(r)}{p_i^2(r) p_f^2(r)} \right)^{1/4} D(r) \\ \times \exp\left\{ \mathrm{i} \left[\frac{t^3 + t'^3}{3} + t\sigma_i(r) + t\sigma_f(r) \right] \right\} \mathrm{d}r \mathrm{d}t \mathrm{d}t'. \end{aligned}$$

The stationary phase point is given by

$$p_i(r_0) = p_i(r_0) = p_0, \tag{8.49}$$

so we can expand σ_α into the neighbourhood of this point

$$\sigma_\alpha(r) = \sigma_\alpha(r_0) + (r - r_0)\, \sigma'_\alpha(r_0) + \dots. \tag{8.50}$$

We shall not detail the calculation, but taking into account the relation $\sigma'_\alpha(r) = p_\alpha(r)/\sqrt{\sigma_\alpha(r)}$, we obtain, without difficulty, the uniform expression of the Franck–Condon factor:

$$F_{if} = \frac{3\pi}{2p_0^2} \frac{[x_i(r_0)x_f(r_0)]^{1/2}}{\left[\frac{3}{2}\left(x_i(r_0) - x_f(r_0)\right)\right]^{1/3}} \times D(r_0) Ai\left\{\left[\frac{3}{2}\left(x_i(r_0) - x_f(r_0)\right)\right]^{2/3}\right\}. \tag{8.51}$$

This result is the same as the one obtained by Bienniek (1977).

With this analytical expression of the Franck–Condon factor, we can now calculate the profile $J(\omega)$. First of all we limit the angular average to the value $\ell_{\max}$, beyond which the penetration at the distance r_0 is no longer classically allowed, $p_0(\ell_{\max}) = 0$. Then we obtain

$$J(\omega) = \int_0^\infty d\varepsilon_i \; e^{-\varepsilon_i/kT} \sum_{\ell=0}^{\ell_{\max}} (2\ell + 1) \left|F_{if}\right|^2. \tag{8.52}$$

Consequently, we can change the discrete sum over the angular momenta into an integration with the help of the Poisson summation formula. [Morse & Feshbach (1953); Berry & Mount (1972)]

$$\sum_{\ell=0}^{\infty} f(\ell) = \frac{1}{h} \sum_{M=-\infty}^{+\infty} e^{-iM\pi} \int_0^\infty f\left[\left(\frac{L}{h}\right) - \left(\frac{1}{2}\right)\right] e^{i2\pi ML/h} dL.$$

Thereafter we shall limit the summation to the first term. In addition, assuming we can expand $p_\alpha(r)$ in the neighbourhood of r_0, we have

$$\begin{cases} x_i(r_0) - x_f(r_0) = \frac{mr_0^2}{2p_0}\Delta V'(r_0) \\ x_i(r_0)x_f(r_0) = r_0^2 p_0^2, \end{cases}$$

where $\delta V'(r_0)$ is the derivative of the difference of the potentials at the Franck–Condon point r_0. The expression (8.51) is now written

$$F_{if} = \frac{3\pi}{2} D(r_0) \left(\frac{4r_0}{3mp_0^2 \Delta V'(r_0)}\right)^{1/3} Ai\left\{\left[\frac{3mr_0^2}{4p_0}\Delta V'(r_0)\right]^{2/3}\right\}. \tag{8.53}$$

Changing the angular quantum number ℓ into the continuous variable $\xi = \ell(\ell+1)$, $J(\omega)$ becomes

$$J(\omega) = \frac{9\pi^2}{4} D^2(r_0) \left(\frac{4r_0}{3mp_0^2 \Delta V'(r_0)}\right)^{2/3} \int_0^\infty d\varepsilon_i \; e^{-\varepsilon_i/kT} \tag{8.54}$$

$$\times \int_0^{\xi_{\max}} d\xi \; p_0^{-4/3}(\xi) Ai^2 \left\{\left[\frac{3mr_0^2 \Delta V'(r_0)}{4p_0(\xi)}\right]^{2/3}\right\},$$

and with the relation

$$\frac{\mathrm{d}p_0(\xi)}{\mathrm{d}\xi} = -\frac{1}{2r_0^2 p_0(\xi)},$$

we obtain

$$J(\omega) = \frac{9\pi^2}{2} D^2(r_0) r_0^2 \left(\frac{4r_0}{3mp_0^2 \Delta V'(r_0)}\right)^{2/3} \int\limits_{V_i(r_0)}^{\infty} \mathrm{d}\varepsilon_i \; \mathrm{e}^{-\varepsilon_i/kT} \tag{8.55}$$

$$\times \int\limits_{0}^{2m[\varepsilon_i - V_i(r_0)]^{1/2}} \mathrm{d}p_0 \; p_0^{-1/3} Ai^2 \left\{ \left[\frac{3mr_0^2 \Delta V'(r_0)}{4p_0}\right]^{2/3} \right\}.$$

For the average over the initial states, the integration is done, not from the zero value of the energy, but from the value $V_i(r_0)$, a value that cancels the upper limit of the other integral. Indeed, the perturbation is unable to reach the point r_0 if it has an energy lower than $V_i(r_0)$, which assumes that the potential curve of the initial state is repulsive.

We apply an integration by parts for the integral on the energies: ε_i

$$\int\limits_{V_i(r_0)}^{\infty} \mathrm{d}\varepsilon_i \; \mathrm{e}^{-\varepsilon_i/kT} \mathfrak{F}(\varepsilon_i) = -kT\mathrm{e}^{-\varepsilon_i/kT} \mathfrak{F}(\varepsilon_i)\Big|_{V_i(r_0)}^{\infty}$$

$$+ \int\limits_{V_i(r_0)}^{\infty} \mathrm{d}\varepsilon_i \; \mathrm{e}^{-\varepsilon_i/kT} \mathfrak{F}'(\varepsilon_i).$$

The first term vanishes because of the exponential for the upper limit and the $\{(\varepsilon_i)$ quantity for the lower limit. The second term eliminates the integration on p_0. Then we obtain for the profile

$$J(\omega) = \frac{27\pi^2}{2} kT \; D^2(r_0) r_0^4 \; \mathrm{e}^{-V_i(r_0)/kT} \mathcal{L}(z), \tag{8.56}$$

where $z = \left(\frac{3}{4} m r_0^2 \Delta V'(r_0)\right)^2 / kT$, and the universal function [Sando & Wormhoudt (1973)]

$$\mathcal{L}(z) = \int\limits_0^{\infty} \mathrm{e}^{-1/t^3} Ai^2(-zt) \frac{\mathrm{d}t}{t^2}. \tag{8.57}$$

This result is valid all along the spectrum. In particular, in the quasi-static approximation, we have to take the asymptotic form of (z):

$$\mathcal{L}(z) \approx \frac{1}{\sqrt{36\pi z}},$$

leading to the quasi-static profile (8.46).

Up to now, we have limited the expansion of $D(r)$ to zero order in the neighbourhood of r_0: $D(r) \approx D(r_0)$. If we now expand $D(r)$ to the second order (cf. Eq. (8.36)), the Franck–Condon amplitude becomes

$$\begin{aligned}\langle\psi_i(r)\,|D(r)|\,\psi_f(r)\rangle &\qquad (8.58)\\ &= \left[D(r_0) - r_0 D'(r_0) + \frac{r_0^2}{2} D''(r_0)\right]\langle\psi_i(r) \mid \psi_f(r)\rangle \\ &\quad + \left[D'(r_0) - r_0 D''(r_0)\right]\langle\psi_i(r)\,|r|\,\psi_f(r)\rangle \\ &\quad + \frac{D''(r_0)}{2}\left\langle\psi_i(r)\,\middle|r^2\middle|\,\psi_f(r)\right\rangle .\end{aligned}$$

As the calculation of the overlap integral $\langle\psi_i(r) \mid \psi_f(r)\rangle$ has already been carried out (formula (8.53)), it only remains to evaluate the integrals $\langle\psi_i(r)\,|r|\,\psi_f(r)\rangle$ and $\left\langle\psi_i(r)\,\middle|r^2\middle|\,\psi_f(r)\right\rangle$. With this intention, we shall again use the uniform wave function expanded in the neighbourhood of the stationary phase point r_0

$$\psi_\alpha(r) = \frac{\sqrt{\pi}}{\sqrt{\sigma'_\alpha(r_0)}} Ai\left[\sigma_\alpha(r_0) + (r - r_0)\sigma'_\alpha(r_0)\right], \quad \alpha = i, f. \qquad (8.59)$$

Let us first calculate the integral

$$\begin{aligned}\langle\psi_i(r)\,|r|\,\psi_f(r)\rangle &\qquad (8.60)\\ &= \frac{\pi}{\left[\sigma'_i(r_0)\sigma'_f(r_0)\right]^{1/2}} \int_{-\infty}^{+\infty} Ai\left[\sigma_i(r_0) + (r - r_0)\sigma'_i(r_0)\right] \\ &\qquad \times Ai\left[\sigma_f(r_0) + (r - r_0)\sigma'_f(r_0)\right].\end{aligned}$$

We find the formula (3.113), where $n = 1$ and

$$\begin{cases} \alpha = \frac{1}{\sigma'_i(r_0)}, & a = \frac{\sigma_i(r_0)}{\sigma'_i(r_0)} - r_0, \\ \beta = \frac{1}{\sigma'_f(r_0)}, & b = \frac{\sigma_f(r_0)}{\sigma'_f(r_0)} - r_0. \end{cases}$$

The formula (3.118) gives the result

$$\langle\psi_i(r)\,|r|\,\psi_f(r)\rangle = A r_0 I_0,$$

where $A = \pi\left[\sigma'_i(r_0)\sigma'_f(r_0)\right]^{-1/2}$ and $I_0 = \langle\psi_i(r) \mid \psi_f(r)\rangle$. So we obtain

$$\langle\psi_i(r)\,|r|\,\psi_f(r)\rangle = r_0\,\langle\psi_i(r) \mid \psi_f(r)\rangle,$$

yielding

$$\langle\psi_i(r)\,|r - r_0|\,\psi_f(r)\rangle = 0. \qquad (8.61)$$

Then the Franck–Condon amplitude (8.58) is limited to

$$\begin{aligned}\langle\psi_i(r)\,|D(r)|\,\psi_f(r)\rangle & \\ &= \left[D(r_0)+\frac{r_0^2}{2}D''(r_0)\right]\langle\psi_i(r)\mid\psi_f(r)\rangle \\ &\quad+\frac{D''(r_0)}{2}\left\langle\psi_i(r)\left|r^2\right|\psi_f(r)\right\rangle.\end{aligned}\tag{8.62}$$

Let us now calculate the term $\left\langle\psi_i(r)\left|r^2\right|\psi_f(r)\right\rangle$. With the same notations as previously, we have

$$\left\langle\psi_i(r)\left|r^2\right|\psi_f(r)\right\rangle = AI_2,$$

where I_2 is defined by the formula (3.119):

$$I_2 = r_0^2 I_0 + \frac{2}{\sigma_i'^3(r_0)-\sigma_f'^3(r_0)}I_0'.$$

We finally obtain the expression of the uniform Franck–Condon amplitude, expanded to the second order

$$\begin{aligned}\langle\psi_i(r)\,|D(r)|\,\psi_f(r)\rangle &= D(r_0)\,\langle\psi_i(r)\mid\psi_f(r)\rangle \\ &\quad+D''(r_0)\frac{\pi}{p_0^4}\left(\frac{3}{2}\right)^{4/3}\frac{[x_i(r_0)x_f(r_0)]^{3/2}}{[x_i(r_0)-x_f(r_0)]^{5/3}} \\ &\quad\times\ Ai'\left\{\left[\frac{3}{2}\left(x_f(r_0)-x_i(r_0)\right)\right]^{2/3}\right\}\end{aligned}\tag{8.63}$$

In particular it can be seen that, in this expression, we no longer have any term in $D'(r_0)$. For a better approximation to the Franck–Condon approximation, it is therefore necessary to expand $D(r)$ up to the second order.

8.3 The semiclassical Wigner distribution

For several years, scientists have sought to establish a correspondence between chaotic classical systems and their quantum correspondent, as well as criteria making it possible to decide whether a quantum system expresses a chaotic behaviour or not. Investigations have focused on phenomena such as irregularity of the spectra [Percival (1973); Berry (1977a)], sensitivity of the spectrum over small disturbances [Pomphrey (1974); Grémaud *et al.* (1993)], distributions of the energy levels [Tabor (1989); McDonald & Kaufmann (1979)], or structures of the stationary states [Heller (1984);

McDonald & Kaufmann (1979)]. However, other authors doubt the existence of quantum chaos [Ford (1989); Ford & Ilg (1992); Ford & Mantica (1992)], and point out that even if quantum systems express a particular behaviour when their traditional equivalent is chaotic, there are serious reasons for thinking that quantum chaos does not exist.

Indeed, it has already been established that finite and closed quantum systems do not express a chaotic behaviour [Ford & Ilg (1992)]. However, the main objection to the existence of quantum chaos remains the linearity of the Schrödinger equation, which makes it insensitive over "the exponential instability of initial conditions" [Ford & Ilg (1992); Dando & Monteiro (1994)], the *sine qua non* condition of chaos. Despite this, it can be objected that any chaotic classical system can be described by the linear Liouville equation, which can be interpreted as a Schrödinger equation [Berry (1991)]. In addition, the criteria for recognising traditional chaos rest upon the concept of trajectory, while the Heisenberg principle of uncertainty excludes the possibility of defining a trajectory in quantum mechanics (see in particular the objections of Ford & Mantica (1992)). If there is no quantum chaos, in the sense of exponential sensitivity to initial conditions, there are several quantum phenomena which reflect the underlying classical chaos. This leads Berry to forge the concept of chaology [Berry (1987), (1989c)]. Berry defined this concept as the study of semiclassical, but nonclassical, phenomena characterized by systems whose classical counterparts exhibit chaos. One method consists in the study of the statistical distribution of energy levels of complex systems. In this way, it has been shown that the energy levels of classically integrable systems distribute according to a Poisson law, whereas the energy levels of chaotic systems follow a statistical law of Wigner type [Berry (1987), Tabor (1989)].

In addition, we lack analytical tools for the study of quantum chaos [Berry (1977a)] and more generally are poorly informed about the correspondence principle between classically chaotic systems and their quantum equivalents, as well as about the quantification of the latter (determination of the eigenvalues and eigenfunctions in terms of classical quantities). However, Berry has shown that the Wigner distribution can be a means of connecting quantum and traditional descriptions for an integrable system (see also Ozorio de Almeida & Hannay (1982); Meredith (1992)).

We thus propose to study here a quantum probability distribution, the Wigner distribution, within classical phase space, in the semiclassical limit $\hbar \to 0$, using the tool of semiclassical methods: the Airy uniform approximation.

8.3.1 *The Weyl–Wigner formalism*

We shall first present the function $W(q,p)$ introduced by Wigner in 1932. This function makes it possible to represent a quantum state $|\psi\rangle$ and is interpreted as the quantum equivalent of a "density" of traditional phase space.

Wigner has shown that, in terms of operators, and in a similar way as in classical mechanics, we have

$$\langle\hat{A}\rangle = \int_{-\infty}^{+\infty} \mathrm{d}^N q \int_{-\infty}^{+\infty} \mathrm{d}^N p\, A(q,p)\, W(q,p),$$

where $W(q,p)$ is the Wigner distribution, defined by

$$W(q,p) = \frac{1}{(\pi\hbar)^N} \int_{-\infty}^{+\infty} \psi(q+\eta)\psi^*(q-\eta)\mathrm{e}^{\mathrm{i}2p\eta/\hbar}\mathrm{d}\eta, \tag{8.64}$$

for a state represented by the wave function $\psi(r)$, here independent of time.

This distribution has many properties (see for example Berry (1977a); O'Connell (1983)). Without entering into great detail, $W(q,p)$ is a real function, but is not always positive. Consequently, $W(q,p)$ cannot be considered as a true density in the phase space. However, it has the usual properties of a probability distribution. Indeed, the Wigner distribution $W(q,p)$ makes it possible to find, by projection, the usual densities of the position space

$$\int_{-\infty}^{+\infty} \mathrm{d}^N p\, W(q,p) = |\langle q \mid \psi\rangle|^2,$$

as well as of the momentum space

$$\int_{-\infty}^{+\infty} \mathrm{d}^N p\, W(q,p) = |\langle p \mid \psi\rangle|^2.$$

So let us note that if $|\psi\rangle$ is normalised to 1, then

$$\int_{-\infty}^{+\infty} \mathrm{d}^N q \int_{-\infty}^{+\infty} \mathrm{d}^N p\, W(q,p) = \langle\psi \mid \psi\rangle = 1.$$

In the semiclassical limit $\hbar \to 0$, if the system is classically integrable, $W(q,p)$ is reduced to the classical density and its equation of motion is reduced to the Liouville equation, which rules the time evolution of a density of classical phase space [Heller (1976)]. Indeed, the semiclassical approach developed by Heller makes it possible to determine the time evolution of a quantum state while following classical trajectories. The link is thus established between the classical field and the quantum one: the wave function is represented by a quantum distribution which moves on a classical path.

If the system considered is not integrable, then the dynamics of this system, as well as the nature of the stationary states and the distribution of the energy levels, are still unknown to us. However, a conjecture was proposed independently by Voros (1976) and Berry (1977a): the microcanonical hypothesis. Even in the case of an ergodic system (i.e. irregular) the Wigner distribution would be reduced, at the classical limit, to the microcanonical distribution That is to say

$$W_m(q,p) = \frac{\delta(E - H(q,p))}{\iint \mathrm{d}p\mathrm{d}q\delta(E - H(q,p))}, \tag{8.65}$$

where δ is the Dirac function.

8.3.2 *The one-dimensional Wigner distribution*

The one-dimensional Wigner distribution $W_1(r,p)$, independent of time, as a function of the position r and the momentum p, is defined by

$$W_1(r,p) = \frac{1}{\pi\hbar} \int_{-\infty}^{+\infty} \psi(r+\eta)\,\psi^*(r-\eta)\,\mathrm{e}^{\mathrm{i}2p\eta/\hbar}\mathrm{d}\eta. \tag{8.66}$$

In this expression we shall replace the wave function $\psi(r)$ of the particle with the mass $m = 1/2$ and the energy E into a potential $V(r)$: $p(r) = \sqrt{E - V(r)}$, by the Airy uniform wave function expanded in the neighbourhood of a turning point r_t (cf. §8.1.3.2, formula (8.27)), i.e.

$$\psi(r) = \frac{\sqrt{\pi}}{\hbar^{1/6}\sqrt{\alpha}} Ai\left[\frac{\alpha}{\hbar^{2/3}}(r - r_t)\right], \tag{8.67}$$

where the constant α is defined by:

$$\alpha^3 = \left[\frac{\mathrm{d}V(r)}{\mathrm{d}r}\right]_{r=r_t}.$$

Note that the definitions of $\psi(r)$ and α are not the same as those given in §8.1.3.2 because we have chosen here to highlight the constant $\hbar$.

It should be pointed out that this method of the turning point cannot be employed in the neighbourhood of a local minimum or maximum of the potential, since in this case we would have $\alpha = 0$. We have now to apply the stationary phase method. The one-dimensional Wigner semiclassical distribution is now written

$$W_1(r,p) = \frac{1}{\hbar^{4/3}} \int_{-\infty}^{+\infty} \frac{1}{\alpha} Ai\left[\frac{\alpha}{\hbar^{2/3}}(r - r_t - \eta)\right] \times Ai\left[\frac{\alpha}{\hbar^{2/3}}(r - r_t + \eta)\right] \mathrm{e}^{\mathrm{i}2p\eta/\hbar}\mathrm{d}\eta. \tag{8.68}$$

The calculation of this integral is given in §3.5.3, formula (3.123), hence the result is

$$W_1(r,p) = \frac{1}{2^{1/3}\hbar^{2/3}\alpha^2} Ai\left\{\frac{2^{2/3}}{\hbar^{2/3}\alpha^2}\left[\alpha^3(r - r_t) + p^2(r)\right]\right\}. \tag{8.69}$$

At the classical limit $\hbar \to 0$, this expression of the one-dimensional, local and semiclassical Wigner distribution must be reduced to the microcanonical distribution. Let us again take the expression (8.69) of the Wigner distribution in the neighbourhood of a turning point r_t and consider the argument of the Airy function, where $\alpha^3 = V'(r_t)$

$$A = (r - r_t)V'(r_t) + p^2,$$

which can be written, by neglecting the higher order terms of the expansion of the potential,

$$A = V(r) - V(r_t) + p^2(r) = (E - H)_{r=r_t},$$

where H is the classical Hamiltonian of the particle with energy E and mass $m = 1/2$, in a potential $V(r)$. The Wigner distribution expanded in the neighbourhood of a turning point hence becomes

$$W_1(r,p) = \frac{1}{2^{1/3}\hbar^{2/3}\alpha^2} Ai\left\{\frac{2^{2/3}}{\hbar^{2/3}\alpha^2}\left[(E - H)_{r=r_t}\right]\right\}.$$

Assuming $\beta = \frac{2^{2/3}}{\hbar^{2/3}\alpha^2}$, we obtain

$$W_1(r,p) = \frac{\beta}{2} Ai\left\{\beta\left[(E - H)_{r=r_t}\right]\right\}. \tag{8.70}$$

The passage to the classical limit is obtained formally by making the Planck constant $\hbar$ tend towards 0, i.e. in making β tend towards infinity. However the Airy function $Ai(x)$ verifies the relation (formula (4.8))

$$\lim_{\beta\to\infty} \beta Ai(\beta x) = \delta(x). \tag{8.71}$$

Therefore, at the classical limit, the Wigner local distribution (8.69) becomes

$$\lim_{\hbar\to 0} W_1(r,p) = \frac{1}{2}\delta(H - E)_{r=r_t}. \tag{8.72}$$

In the semiclassical limit, the Wigner distribution (8.69) expanded in the neighbourhood of a turning point is reduced, as it should be, to the microcanonical distribution.

Note that the factor 1/2 results from the fact that by carrying out the expansion in the neighbourhood of a turning point, we have taken into account only half of the phase space.

8.3.3 *The two-dimensional Wigner distribution*

In order to determine the analytical expression of the two-dimensional Wigner distribution, we shall proceed in the same way as in the previous section, i.e. we define the quantities $\sigma(x,y) = \left(\frac{3}{2}\int_{x_t}^{x} p_1(u,y)\mathrm{d}u\right)^{2/3}$ and $\xi(x,y) = \left(\frac{3}{2}\int_{y_t}^{y} p_2(x,u)\mathrm{d}u\right)^{2/3}$, where p_1 and p_2 are defined by separating the momentum in two parts, such that $p_1^2 + p_2^2 = E - V$.[2] The two-dimensional uniform wave function can thus be written

$$\psi(x,y) = \frac{\pi}{\hbar^{1/3}\left[\sigma_x'(x,y)\xi_y'(x,y)\right]^{1/2}} Ai\left[\frac{\xi(x,y)}{\hbar^{2/3}}\right] Ai\left[\frac{\sigma(x,y)}{\hbar^{2/3}}\right]. \tag{8.73}$$

As previously, we expand this wave function in the neighbourhood of a turning point, $r_t = (x_t, y_t)$, to obtain

$$\begin{aligned}\psi(x,y) = \frac{\pi}{\hbar^{1/3}\sqrt{ad}} & Ai\left[\frac{a}{\hbar^{2/3}}(x-x_t) + \frac{b}{\hbar^{2/3}}(y-y_t)\right] \\ & \times Ai\left[\frac{c}{\hbar^{2/3}}(x-x_t) + \frac{d}{\hbar^{2/3}}(y-y_t)\right],\end{aligned} \tag{8.74}$$

where a, b, c and d are the two-dimensional equivalents of the previously defined constant α, leading to

$$\begin{cases} a = \left[\frac{\mathrm{d}\sigma(x,y)}{\mathrm{d}x}\right]_{(x,y)=(x_t,y_t)} & b = \left[\frac{\mathrm{d}\sigma(x,y)}{\mathrm{d}y}\right]_{(x,y)=(x_t,y_t)} \\ c = \left[\frac{\mathrm{d}\xi(x,y)}{\mathrm{d}x}\right]_{(x,y)=(x_t,y_t)} & d = \left[\frac{\mathrm{d}\xi(x,y)}{\mathrm{d}y}\right]_{(x,y)=(x_t,y_t)} \end{cases}.$$

A rather long, though not particularly difficult, calculation leads to the analytical expression of the two-dimensional Wigner semiclassical distribution,

[2] This separation in p_1 and p_2 can appear arbitrary, but it is usual to distinguish one of the variables compared to the other, while defining for example $p_1(x)$ and $p_2(x,y)$ [Martens *et al.* (1988)].

expanded in the neighbourhood of a turning point (x_t, y_t)

$$W_2(x, y, p_x, p_y) \tag{8.75}$$
$$= \frac{1}{2^{2/3}\hbar^{4/3}a^2d^2} Ai\left\{\frac{2^{2/3}}{\hbar^{2/3}}\left[(ax+by)+\left(\frac{p_x d - p_y c}{\Delta}\right)^2\right]\right\}$$
$$\times Ai\left\{\frac{2^{2/3}}{\hbar^{2/3}}\left[(cx+dy)+\left(\frac{p_x b - p_y a}{\Delta}\right)^2\right]\right\},$$

where we posed, to reduce the notation and without loss of generality, $x = x - x_t$ and $y = y - y_t$.

Note that, as in the one-dimensional case, this method is not applicable in the neighbourhood of a local minimum or maximum of the potential $(\delta = 0)$.

Let us first examine the case where the Hamiltonian variables are separable, i.e. where the Hamiltonian can be written in the form $H(x, y) = h_1(x) + h_2(y)$. The functions $\xi(x, y)$ and $\sigma(x, y)$ are then reduced to

$$\begin{cases} \sigma(x,y) = \sigma(x) \\ \xi(x,y) = \xi(y). \end{cases}$$

We thus have, in this case, $b = c = 0$ (cf. formula (8.3.3)) and the discriminant δ is reduced to $\delta = ad$. Consequently, the expression (8.75) of the Wigner distribution becomes

$$W_2^{\text{sep}}(x, p_x; y, p_y) \tag{8.76}$$
$$= \frac{1}{2^{2/3}\hbar^{4/3}a^2d^2} Ai\left\{\frac{2^{2/3}}{\hbar^{2/3}}\left[ax+\left(\frac{p_x}{a}\right)^2\right]\right\}$$
$$\times Ai\left\{\frac{2^{2/3}}{\hbar^{2/3}}\left[dy+\left(\frac{p_y}{d}\right)^2\right]\right\}.$$

Thus, in the case of the separable variables (integrable case), W_2^{sep} is reduced to the product of two one-dimensional Wigner distributions and consequently, at the classical limit, the two-dimensional semiclassical Wigner distribution expanded in the neighbourhood of a turning point is reduced again to the microcanonical distribution.

Let us return now to the more general case where the variables are not *a priori* separable. The two-dimensional Wigner distribution is then given by the expression (8.75). We carry out the linear changes of variable

$$\begin{cases} X = ax + by \\ Y = cx + dy\,. \end{cases}$$

As the discriminant does not cancel ($\Delta = ad - bc \neq 0$), the inverse transformation is

$$\begin{cases} x = \frac{1}{\Delta}(dX - bY) \\ y = \frac{1}{\Delta}(-cX + aY) \, . \end{cases}$$

However the conjugate variables x and p_x verify the relation $p_x = \frac{\partial L}{\partial \dot{x}}$, where $L(\dot{x}, x, \dot{y}, y)$ is the Lagrangian of the system. After transformation we shall thus have the new combined moment

$$P_X = \frac{\partial L}{\partial \dot{X}} = \frac{\partial L}{\partial \dot{x}} \frac{\partial \dot{x}}{\partial \dot{X}} + \frac{\partial L}{\partial \dot{y}} \frac{\partial \dot{y}}{\partial \dot{Y}}$$

i.e.

$$P_X = p_x \frac{d}{\Delta} + p_y \frac{-c}{\Delta}. \tag{8.77}$$

Similarly, we have

$$P_Y = p_x \frac{-b}{\Delta} + p_y \frac{a}{\Delta}. \tag{8.78}$$

The expression of the Wigner distribution then becomes

$$W_2(X, P_X; Y, P_Y) = \frac{1}{2^{2/3} \hbar^{4/3} a^2 d^2} Ai \left\{ \frac{2^{2/3}}{\hbar^{2/3}} \left(X + P_X^2 \right) \right\} \times Ai \left\{ \frac{2^{2/3}}{\hbar^{2/3}} \left(Y + P_Y^2 \right) \right\}. \tag{8.79}$$

The linear transformation (8.3.3) is thus a canonical transformation, which transforms the Hamiltonian $H(x, y, p_x, p_y)$ into the Hamiltonian $\mathcal{H}(X, P_X; Y, P_Y)$. As previously, we can therefore pass to the classical limit and write

$$W_2(X, P_X; Y, P_Y) = \frac{1}{4} \delta \left(\mathcal{H} - \mathcal{E} \right)_{X_t, Y_t} .$$

Thus, the reverse linear transformation (8.3.3) makes it possible, at least locally, to obtain the microcanonical distribution

$$\delta \left(H(x, y, p_x, p_y) - E \right). \tag{8.80}$$

Even if we are not in the case of a system whose variables are separable, the fact of expanding the function in the neighbourhood of a turning point, i.e. of linearising the potential in this neighbourhood, allows, by means of a linear and canonical transformation, the problem to be reduced to the case of separable variables.

8.3.4 *Configuration of the Wigner distribution in the phase space*

Since we now have an analytical expression of the Wigner distribution, it is natural to represent it in the phase space. On the one hand we already know, at least qualitatively, the behaviour of this function for an integrable system [Berry (1977b)]. On the other hand, we are unaware of its behaviour in the case of a non-integrable system. However, Berry put forth the hypothesis that the Wigner distribution would be like a series of minima and maxima randomly distributed in the phase space. Figure (8.8) shows a cut of the phase space for an integrable system, and Fig. (8.9) for a non-integrable system. We also give, in the first case, the configuration of the Wigner distribution, and in the second one the distribution of Berry's conjecture. The boxes illustrate the fact that each time only half of the phase space is considered.

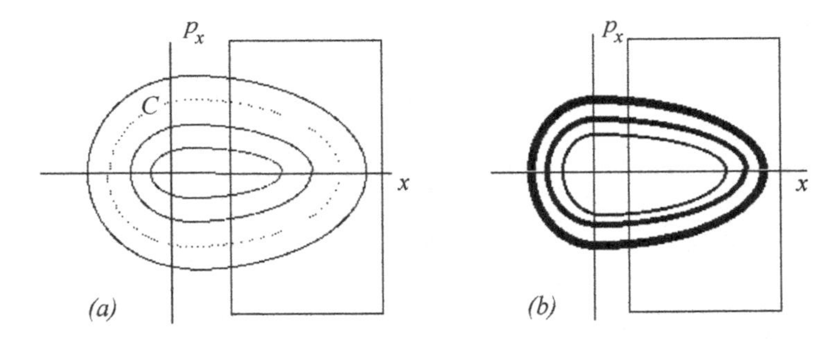

Fig. 8.8 (a): Section S of the phase space (classical case) for a particle in a bound state energy E_m. The curve C is, in classical terms, the intersection of the torus I_m with S. (b): The corresponding Wigner distribution is concentrated in the neighbourhood of C by forming fringes. (Berry, 1977b).

The expression of the Wigner distribution given by the formula (8.79), except for the constants, can be written

$$W_2 = Ai\left(X + P_X^2\right). \tag{8.81}$$

This function is represented on Fig. (8.10), in the phase space (x, p), reduced here to (x, p_x).

As already mentioned, we expanded the distribution in the neighbourhood of only one turning point, consequently Fig. (8.10) represents only half of the distribution.

As can be seen on this figure, we obtain a regular distribution, identical to Fig. (8.8), having the shape Berry called "Airy fringes".

The expression to the two-dimensional Wigner distribution given by the formula (8.75) lends itself particularly well to our aim. It consists of replacing the constants a, b, c and d with the elements of a unimodular transformation matrix.

It is *a priori* possible to choose any matrix whose determinant is 1, in order to ensure the conservation of the volume of the phase space. Indeed, the volume occupied by a dynamical system in the phase space remains unchanged during the evolution of this system. We shall, therefore, by means of a transformation matrix, perturb the system by locally deforming the space and observe the reaction on the distribution of these constraints.

Among the unimodular transformation matrices, we consider the shearing matrix

$$\begin{bmatrix} 1 & 0 \\ \gamma & 1 \end{bmatrix},$$

of which the deformation effects are more sensitive than with the rotation or hyperbolic rotation matrices. For various values of the parameter γ, we shall represent the Wigner distribution in the plane (x, p_x) of the phase space. The formula (8.75) gives the expression (except for the constants)

$$W_2 = Ai\left[x + p_x^2\right] Ai\left[\gamma x\right]. \tag{8.82}$$

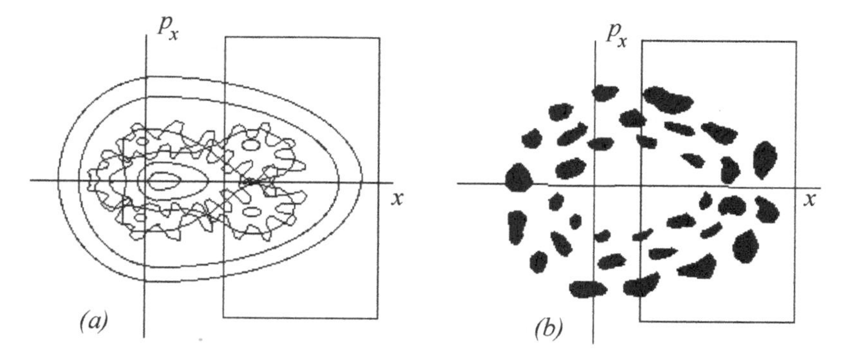

Fig. 8.9 (a): Section corresponding to Fig. 8.8, with a non-integrable perturbation. Some tori are still present, but irregular trajectories appear. (b): The corresponding Wigner distribution in a state where classical motion would be irregular. W would take the form of a series of random minima and maxima covering the areas occupied by these irregular trajectories [Berry, (1977b)].

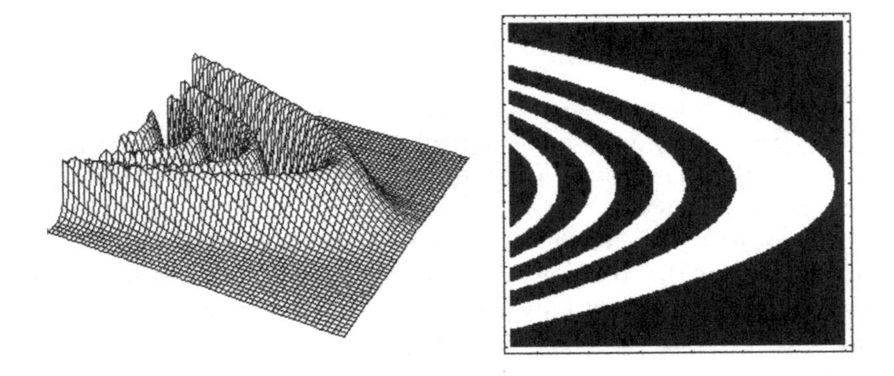

Fig. 8.10 The two-dimensional Wigner distribution, in the case of separable variables, represented in the plane defined by (X, P_X). On the right-hand side, a cut of the distribution showing regular minima and maxima is given.

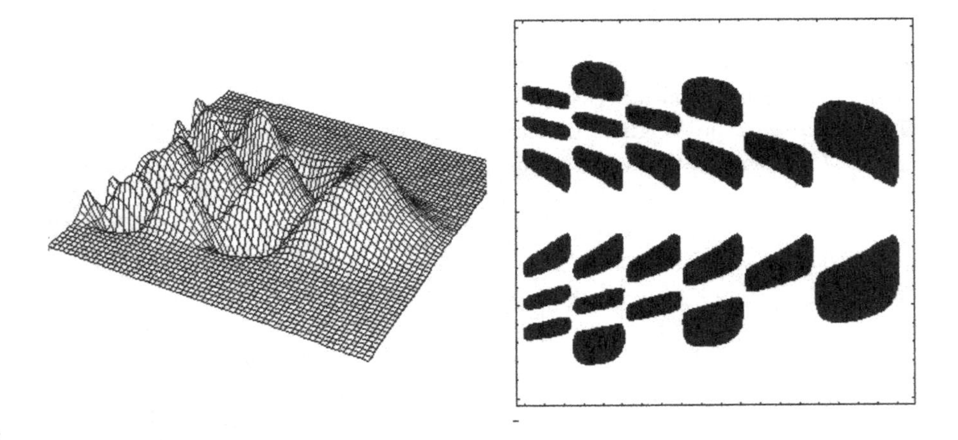

Fig. 8.11 The two-dimensional Wigner distribution ($\gamma = 1$) in the (x, p_x) plane. On the right the cut of the distribution shows the distribution of peaks formed by minima and maxima.

For $\gamma = 0$, W_2 is reduced to the one-dimensional distribution in the (x, p_x) plane. For other values of the parameter γ, the distribution undergoes very significant deformations. In particular, for the value $\gamma = 1$, we observe the distribution of the minima and maxima of W_2 in Fig. (8.11). This distribution, in qualitative terms, resembles Berry's conjecture, on Fig. (8.9), where he assumed that the minima and maxima of the semiclassical distribution are randomly distributed in the ergodic system case. However, we obtain a qualitatively similar distribution though we are in an integrable case since the variables are separable.

It would therefore seem that it is not necessary to be in an ergodic system, leading to a relatively irregular form for the Wigner distribution. Note that Fig. (8.11) still seems "regular" enough, but we are in the particular case of an expansion in the neighbourhood of a turning point. However, even within the framework of a very simple model, where the variables are separable and the trajectories regular, the distribution can have no simple structure, at least locally, which seems to invalidate the reciprocal of Berry's conjecture.

Recently Tomsovic & Heller (1993) built "chaotic" wave functions i.e. the wave functions of a system in which the dynamics of the corresponding classical system are chaotic, from purely semiclassical methods. Their results were applied to the "stadium billiard problem" (classically chaotic) and stationary states were obtained, which seem to be the same for the Wigner distribution as the Berry conjecture. In other words, the same stionary states were observed as those obtained by semiclassical methods, but where all possibility of chaotic behaviour is excluded.

8.4 Airy transform of the Schrödinger equation

A few years ago, Berry & Balazs (1979) built a wave packet which moves without dispersion and which accelerates, without being subjected to an external force, thus seeming to contradict the Ehrenfest theorem. Let us consider the Schrödinger equation for a free particle with a mass m

$$\mathrm{i}\hbar\frac{\partial\psi}{\partial t} = -\frac{\hbar^2}{2m}\frac{\partial^2\psi}{\partial x^2}, \tag{8.83}$$

and the wave packet defined at $t = 0$ by the relation $\psi(x, 0) = Ai(Bx/\hbar^{2/3})$, where B is an arbitrary positive constant. The solution to Eq. (8.83) is the *a priori* strange wave packet

$$\psi(x,t) = \exp\left[i\left(\frac{B^3t}{2m\hbar}\right)\left(x - \frac{B^3t^2}{6m^2}\right)\right] Ai\left[\frac{B}{\hbar^{2/3}}\left(x - \frac{B^3t^2}{4m^2}\right)\right], \tag{8.84}$$

for it is apparently in contradiction with quantum mechanics. Indeed, this "Airy wave packet" (or more exactly the probability density $|\psi(x,t)|^2$) does not disperse with time and, moreover, accelerates without any force acting on it.

However, Berry & Balazs show that this solution satisfies the Ehrenfest theorem, because the Airy wave packet does not have a centre of mass that is perfectly localised. Indeed, the Airy function $Ai(x)$ is not square integrable. The wave packet (8.84) cannot thus represent the density of probability corresponding to a single particle. Consequently, the Airy wave packet represents a set of particles from an infinite number, i.e. a family of semiclassical orbits in the phase space (this is the analogue of the plane wave for the theory of diffraction).

Greenberger (1980) has shown, by a Galilean transformation dependent on time, that this solution is equivalent to the stationary state of a particle in a uniform gravitational potential. The system in free fall is described by the same type of function, except for argument and phase shift.

Recent work [Besieris *et al.* (1994); Nassar *et al.* (1995)] shows that a change of variables depending on time can lead to alternatives of the Airy wave packet of (8.84), in more general cases than that of a null or constant force. We can thus build dispersive stationary wave packets, or wave packets which do not accelerate if not subjected to an external force....

We shall show, using the Airy transform (cf. §4.2), that the connection between the Schrödinger equation for the free particle and the one for free fall does not require a transformation that depends on time. Let us consider the Schrödinger equation, the atomic units ($\hbar = m = e = 1$) will be used

$$\mathrm{i}\frac{\partial \psi}{\partial t} = -\frac{1}{2}\frac{\partial^2 \psi}{\partial x^2} + V(x)\psi. \tag{8.85}$$

The Airy transform of this equation is

$$\mathrm{i}\frac{\partial \varphi_a}{\partial t} = -\frac{1}{2}\frac{\partial^2 \varphi_a}{\partial y^2} + \int W_a(y,z)\varphi_a(z)\mathrm{d}z, \tag{8.86}$$

where $\varphi_a(y,t)$ is the Airy transform of $\psi(x,t)$, and where the potential $V(x)$ is transformed into a non local potential $W_a(y,z)$

$$W_a(y,z) = \frac{1}{a^2}\int V(x) Ai\left(\frac{y-x}{a}\right) Ai\left(\frac{z-x}{a}\right)\mathrm{d}x. \tag{8.87}$$

Let us examine some particular cases. If the potential $V(x)$ is null or constant, we can see, according to Eqs. (8.85) and (8.86), that the Schrödinger equation is invariant under the Airy transform. Now let us consider the linear potential $V(x) = Fx$. Thanks to the properties of the Airy transform, the Schrödinger equation (8.85) is written into the self-similar form

$$\mathrm{i}\frac{\partial \varphi_a}{\partial t} = -\left(\frac{1}{2} + Fa^3\right)\frac{\partial^2 \varphi_a}{\partial y^2} + Fy\varphi_a(y). \tag{8.88}$$

Thus a scaling of space and time leaves the equation invariant. Choosing as parameter $a = -(1/2F)^{1/3}$ in Eq. (8.88), we obtain the simple solution

$$\varphi_a(y,t) = \mathrm{e}^{-\mathrm{i}tFy}\phi(y),$$

where $\phi(y)$ is the Airy transform of the initial wave function. The interpretation of this result is that the Airy transform neutralises the effect of the uniform field in such a way that, except for a phase factor, the wave function remains the same as at initial time. Assuming that $\phi(y)$ is square integrable, the Plancherel–Parseval rule gives us

$$\int |\phi|^2 \mathrm{d}y = \int |\varphi_a|^2 \mathrm{d}y = \int |\psi|^2 \mathrm{d}x < \infty,$$

i.e. the Airy inverse transform is also square integrable. We can then calculate explicitly the wave function

$$\psi(x,t) = \frac{1}{2\pi}\int \exp\left[\mathrm{i}\left(\frac{\xi^3}{6F} + \xi x\right)\right]\hat{\varphi}\left(\xi + Ft\right)\mathrm{d}\xi,$$

where $\hat{\varphi}$ is the Fourier transform of φ.

So now considering the solution of null energy as initial wave function $\hat{\varphi}(\xi) = \mathrm{e}^{\mathrm{i}\xi^3/6F}$, which is not square integrable, we obtain the wave packet

$$\psi(x,t) = \mathrm{e}^{-\mathrm{i}\frac{1}{2}Ftx} Ai\left[F^{1/3}\left(x + \frac{Ft^2}{4}\right)\right], \tag{8.89}$$

which is the solution of the Schrödinger equation for a particle in free fall. We can compare this solution with

$$\psi_0(x,t) = \mathrm{e}^{\mathrm{i}\left(\frac{F^2t^3}{12} + \frac{1}{2}Ftx\right)} Ai\left[F^{1/3}\left(x - \frac{Ft^2}{4}\right)\right], \tag{8.90}$$

which is nothing but Eq. (8.84) where $m = \hbar = 1$ and $F = B^3$. The only difference (except for the phase factor containing the term in t^3) between the latter two equations is the direction of propagation: the relation (8.89) describes a backwards propagation and Eq. (8.90) a forwards propagation.

Exercises

(1) With an appropriate change of variable and function, find the general solution of the Schrödinger equation in a time-dependent electric field

$$\mathrm{i}\frac{\partial\psi}{\partial t} = -\frac{1}{2}\frac{\partial^2\psi}{\partial x^2} + F(t)\,x\psi.$$

Hint: See the paper by [Feng (2001)].

(2) Find the energy eigenvalues of a particle in a one-dimensional box submitted to an electric field $V(x) = Fx$. See Fig. (8.12-a).

(3) Quantum wells formed in semiconductor heterostructures have received considerable theoretical consideration for their optical device applications. A simple model of tunneling in such devices is the study of the energy spectrum of a particle in a square well submitted to an electric field $V(x) = Fx$. See Fig. (8.12-b). *Hint*: For further discussion, see the following paper [Panda & Panda (2001)].

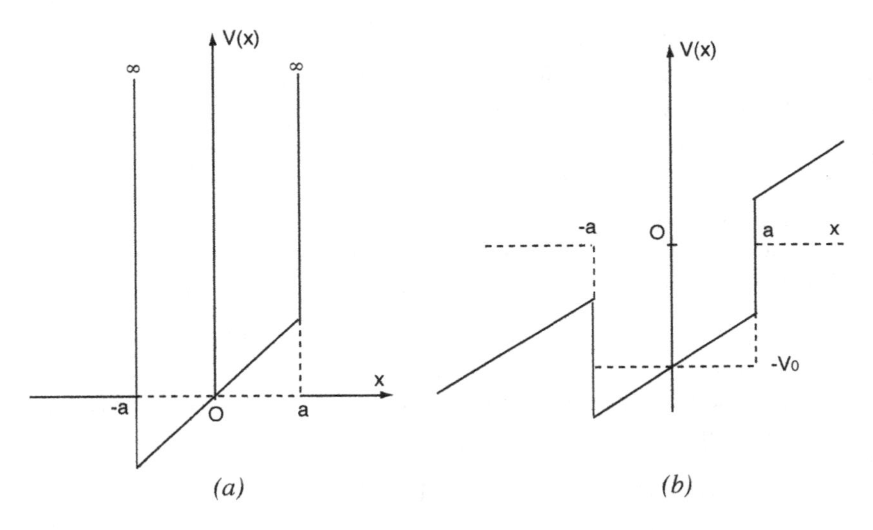

Fig. 8.12 (a) Particle in a box submitted to an electric field. (b) Particle in a square well submitted to an electric field.

Appendix A

Numerical Computation of the Airy Functions

A.1 Homogeneous functions

Except for using the numerical tables of Miller (1946) and Abramowitz & Stegun (1965), the calculation of the Airy functions can be carried out by means of the algorithms of Gordon (1969), (1970) and Lee (1980). Indeed, if we use a Gauss method of quadrature, the Airy function $Ai(x)$ should be replaced by the following sums (here we highlight the formulae (2.33) and (2.35)), where $\xi = \frac{2}{3}x^{3/2}$, $x > 0$

$$Ai(x) \approx \frac{e^{-\xi}}{2\pi^{1/2}x^{1/4}} \sum_{i=1}^{n} \frac{w_i}{1+\left(\frac{x_i}{\xi}\right)} \tag{A.1}$$

$$Ai(-x) \approx \frac{1}{\pi^{1/2}x^{1/4}} \sum_{i=1}^{n} w_i \frac{\cos\left(\xi-\frac{\pi}{4}\right)+\frac{x_i}{\xi}\sin\left(\xi-\frac{\pi}{4}\right)}{1+\left(\frac{x_i}{\xi}\right)^2}, \tag{A.2}$$

and the $Bi(x)$ function by (formulae (2.34) and (2.36))

$$Bi(x) \approx \frac{e^{\xi}}{\pi^{1/2}x^{1/4}} \sum_{i=1}^{n} \frac{w_i}{1-\left(\frac{x_i}{\xi}\right)} \tag{A.3}$$

$$Bi(-x) \approx \frac{1}{\pi^{1/2}x^{1/4}} \sum_{i=1}^{n} w_i \frac{\sin\left(\xi-\frac{\pi}{4}\right)-\frac{x_i}{\xi}\cos\left(\xi-\frac{\pi}{4}\right)}{1+\left(\frac{x_i}{\xi}\right)^2}. \tag{A.4}$$

w_i are the weight factors corresponding to the integration points x_i [Gordon (1970)]. In Table (A.1), we give the partition of $\mathbb{R}$ for the calculation of the homogeneous Airy functions. We give in Table (A.2) the values of w_i and x_i for integration in 10 points. In all the cases, the neighbourhood of the origin will be calculated with the help of the ascending series, given by the formulae (2.38) and (2.39).

Table A.1 Partition of $\mathbb{R}$ for the calculation of the homogeneous Airy functions

x	$Ai(x)$	$Bi(x)$
$x < -3.7$	integration A.2	integration A.4
$-3.7 < x < 2.35$	series 2.38	series 2.39
$2.35 < x < 8.5$	integration A.1	series 2.39
$8.5 < x$	integration A.1	integration A.3

Table A.2 Weight factors w_i and integration abscissas x_i for the Gauss quadrature method with 10 points.

	x_i	w_i
1	1.408308107218096D + 01	3.154251576296478D − 14
2	1.021488547919733D + 01	6.639421081958493D − 11
3	7.441601845045093	1.758388906134567D − 08
4	5.307094306178192	1.371239237043582D − 06
5	3.634013502913246	4.435096663928435D − 05
6	2.331065230305245	7.155501091771825D − 04
7	1.344797082460927	6.488956610333538D − 03
8	6.418885836956729D − 01	3.644041587577328D − 02
9	2.010034599812105D − 01	1.439979241859100D − 01
10	8.059435917205284D − 03	8.123114133626148D − 01

It should be noted that, for the large values of the argument of Airy functions ($|x| > 10$), we can carry out the Gauss quadrature with fewer integration points and the use of asymptotic series (cf. §2.1.4.3), without any loss of precision. We can easily test the validity of these various methods, in particular at the points of connection, with the help of the Wronskian of the Airy functions (formula (2.6))

$$Ai(x)Bi'(x) - Ai'(x)Bi(x) = \frac{1}{\pi}.$$

There are other methods to numerically calculate the Airy functions. Among them, Krüger (1981), proposed the following method for the calculation of $Ai(x)$. Let us consider the definition of the Airy function

$$Ai(x) = \frac{1}{2\pi} \int_{-\infty}^{+\infty} \mathrm{e}^{\mathrm{i}\left(t^3/3+tx\right)} \mathrm{d}t.$$

By carrying out the change of variable $t \to t + \mathrm{i}\alpha$, in the complex plane, we obtain

$$Ai(x) = \frac{1}{2\pi} \mathrm{e}^{\alpha^3/3 - \alpha x} \int_{-\infty + \mathrm{i}\alpha}^{+\infty + \mathrm{i}\alpha} \mathrm{e}^{-\alpha t^2 + \mathrm{i}\left[t^3/3 + \left(x - \alpha^2\right)t\right]} \mathrm{d}t.$$

We can then use the Poisson summation formula to get

$$Ai(x) = h \sum_{k=0}^{\infty} \psi \left[h \left(n + \frac{1}{2} \right) \right] - \Delta(p),\ p = \frac{2\pi}{h},\ h > 0, \tag{A.5}$$

and

$$\psi(t) = \frac{1}{\pi} \mathrm{e}^{\alpha^3/3 - \alpha x - \alpha t^2} \cos \left[\frac{t^3}{3} + t\left(x - \alpha^2 1\right) \right],$$

$$\Delta(p) = \sum_{r=1}^{\infty} (-1)^r R(rp),$$

$$R(p) = R(-p) = \mathrm{e}^{\alpha p} Ai(x+p) + \mathrm{e}^{-\alpha p} Ai(x-p).$$

The expression (A.5) is an exact trapezoidal summation formula. Under some conditions, detailed by Krüger (1981), we can neglect the term $\Delta(p)$, and limit the summation to a range N, so that we obtain for $\alpha = 1$

$$Ai(x) \approx h \sum_{k=0}^{N} \psi \left[h \left(k + \frac{1}{2} \right) \right]$$

where

$$\psi(t) = \frac{1}{\pi} \mathrm{e}^{-x - t^2 + 1/3} \cos \left[\frac{t^3}{3} + t(x-1) \right].$$

If we choose $h = \frac{2\pi}{26}$ and $N = 24$, these two formulae give the value of $Ai(x)$ for $-10 < x < 0$ with a relative error lower than 3.10^{-8}, and for $x > 0$, an error of about 10^{-12}.

A.2 Inhomogeneous functions

As previously, the functions $Gi(x)$ and $Hi(x)$ can be calculated using the sums, for $x > 0$ (cf. formulae (2.138) and (2.128))

$$Gi\,(x) = -\frac{1}{\pi} \sum_{i=1}^{n} w_i \mathrm{e}^{-x_i x/2} \cos \left(\frac{\sqrt{3}}{2} x_i x + \frac{2\pi}{3} \right) \tag{A.6}$$

$$Hi\,(-x) = \frac{1}{\pi} \sum_{i=1}^{n} w_i \mathrm{e}^{-x_i x}. \tag{A.7}$$

$Gi(-x)$ and $Hi(x)$, result from the previous two equations and from formulae (A.3) and (A.4) thanks to the relation between Bi, Gi and Hi (formula (2.131)) $Bi(x) = Gi(x) + Hi(x)$. The neighbourhood of the origin will be calculated by using the ascending series defined in §2.3.3: formulae (2.141) for Hi, (2.39) (series of Bi) and the relation (2.131) for $Gi(x)$.

In Table (A.3), we give the partition of $\mathbb{R}$ for the calculation of the inhomogeneous Airy functions, and in Table (A.4) the weight factors of integration w_i and the corresponding abscissae x_i, calculated thanks to the algorithm of Gordon (1968).

Exercises

Check your favourite computer library for the computation of Airy functions with the following tests:

(1) Check the Wronskian relationships
$W\{Ai(x),\ Bi(x)\} = \frac{1}{\pi}$ and
$W\left\{Ai^2(x), Ai(x)Bi(x), Bi^2(x)\right\} = \frac{2}{\pi^3}$.

(2) Compare the logarithmic derivative of Airy functions with a numerical solution to the Riccati equation $u' + u^2 = x$.

(3) In the complex plane: check the relation $Ai(x)+\mathrm{j}Ai(\mathrm{j}x)+\mathrm{j}^2Ai(\mathrm{j}^2x) = 0$, and the corresponding one for Bi. Check the relation (2.165) of §2.4.2.

Table A.3 Partition of $\mathbb{R}$ for the calculation of the Airy inhomogeneous functions.

x	$Gi(x)$	$Hi(x)$
$x < -6$	A.7 and 2.131	integration A.7
$-6 < x < -4$	integration A.7 and 2.131	integration A.7
$-4 < x < 6$	series 2.141 and 2.131	series 2.141
$6 < x < 8$	integration A.6	series 2.141
$8 < x$	integration A.6	A.6 and 2.131

Table A.4 Weight factors w_i and integration abscissas x_i for the Gauss method with 15 points.

	x_i	w_i
1	1.4576978176136D − 02	3.8053986078615D − 02
2	8.1026698767654D − 02	9.6220284128805D − 02
3	2.0814345959022D − 01	1.5721761605002D − 01
4	3.9448412556694D − 01	2.0918953325833D − 01
5	6.3156478398822D − 01	2.3779904013329D − 01
6	9.0760339986136D − 01	2.2713825749406D − 01
7	1.2106768087608	1.7328458073252D − 01
8	1.5309839772429	9.869554247686D − 02
9	1.8618445873124	3.893631493517D − 02
10	2.1997121656815	9.812496327697D − 03
11	2.5438398040282	1.4391914183288D − 03
12	2.8961730431054	1.0889100255168D − 04
13	3.2620667311773	3.5468667194632D − 06
14	3.6533718875065	3.5907188198098D − 08
15	4.1023767739755	5.1126116783291D − 11

Bibliography

Ablowitz, M.J. and Segur, H. (1977). Exact linearisation of a Painlevé transcendent, *Phys. Rev. Lett.* **38**, pp. 1103–1106.

Ablowitz, M.J., Ramani, A. and Segur, H. (1980). A connection between nonlinear evolution equations and ordinary differential equation of P-type I, *J. Math. Phys.* **21**, pp. 715–721.

Ablowitz, M.J. and Segur, H. (1981). Solitons and the inverse scattering transform, SIAM, Philadelphia.

Ablowitz, M.J. and Clarkson P.A. (1991). Solitons, Nonlinear evolution equations and inverse scattering transform, *Lecture Note Series # 149*, Cambridge University Press.

Abramowitz, M. and Stegun, I. (1965). Handbook of Mathematical Functions, Dover Publications, New York.

Airy, G.B. (1838). On the intensity of light in the neighbourhood of a caustic, *Trans. Camb. Phil. Soc.* **6**, pp. 379–401.

Airy, G.B. (1845). Tides and waves, Encyclopedia Metropolitana **5**, pp. 241–396.

Airy, G.B. (1849). Supplement to a paper "On the intensity of light in the neighbourhood of a caustic", *Trans. Camb. Phil. Soc.* **8**, pp. 595–599.

Airy, W. (1896). Autobiography of G.B. Airy, W. Airy editor.

Albright, J.R. (1977). Integrals of products of Airy functions, *J. Phys. A* **10**, pp. 485–490.

Albright, J.R. and Gavathas, E.P. (1986). Integrals involving Airy functions, *J. Phys. A* **19**, pp. 2663–2665.

Alexander, M.H. and Manolopoulos, D.E. (1987). A stable linear reference potential algorithm for solution of the quantum close-coupled equations in molecular scattering theory, *J. Chem. Phys.* **86**, pp. 2044–2050.

Apelblat, A. (1980). Mass transfer with a chemical reaction of the first order: analytical solutions, *The Chemical Engineering Journal* **19**, pp. 19–37.

Apelblat, A. (1982). Mass transfer with a chemical reaction of the first order. Effect of axial diffusion, *The Chemical Engineering Journal* **23**, pp. 193–203.

Apelblat, A. (1996). Tables of integrals and series, Verlag Harri Deutsch, Thun and Frankfurt am Main.

Arnold, V.I. (1981). Singularity Theory (Selected papers). Cambridge University Press, Cambridge.

Arrighini, G.P., Durante, N. and Guidotti, C. (1999). The method of Airy averaging and some useful applications, *J. Math. Chem.* **25**, pp. 93-103.

Aspnes, D.E. (1966). Electric-field effects on optical absorption near thresholds in solids, *Phys. Rev.* **147**, pp. 554–566.

Aspnes, D.E. (1967). Electric-field effects on the dielectric constant of solids, *Phys. Rev.* **153**, pp. 972–982.

Balazs, N.L. and Zipfel, G.G. (1973). Quantum oscillations in the semiclassical fermion μ-space density, *Ann. Phys.* **77**, pp. 139–156.

Banderier, C., Flajolet, P., Schaeffer, G. and Soria, M. (2000). Planar maps and Airy phenomena, *Proceeding of ICALP 2000; U. Montanari et al. (Eds.)* Springer Verlag, pp. 388–402.

Baldwin, P. (1985). Zeroes of generalized Airy functions, *Mathematika* **32**, pp. 104–117.

Basor, E.L. and Widom, H. (1999). Determinants of Airy operators and applications to random matrices, *J. Statis. Phys.* **96**, pp. 1–20.

Berry, M.V. and Mount, K.E. (1972). Semiclassical approximations in wave mechanics, *Rep. Prog. Phys.* **35**, pp. 315–397.

Berry, M.V. (1976). Waves and Thom's theorem. *Adv. Phys.* **25**, pp. 1–26.

Berry, M.V. (1977a). Semi-classical mechanics in phase space : a study of Wigner's function, *Phil. Trans. R. Soc. London A* **287**, pp. 237–271.

Berry, M.V. (1977b). Regular and irregular semiclassical wavefunctions, *J. Phys.* A **10**, pp. 2083–2091.

Berry, M.V. and Balazs, N.L. (1979a). Nonspreading wave packets, *Am. J. Phys.* **47**, pp. 264–267.

Berry, M.V., Nye J.F. and Wright, F.J. (1979b). The elliptic umbilic diffraction catastrophe, *Phil. Trans.R. Soc. A* **291**, pp. 453–484.

Berry, M.V. and Wright, F.J. (1980a). Phase-space projection identities for diffraction catastrophes, *J. Phys. A* **13**, pp. 149–160.

Berry, M.V. and Upstill, C. (1980b). Catastrophe optics: morphologies of caustics and their diffraction patterns. *Progress in Optics XVIII* pp. 257–346 (and references therein).

Berry, M.V. (1987). The Bakerian Lecture 1987: Quantum chaology. *Proc. R. Soc. Lond. A* **413**, pp. 183–198.

Berry, M.V. (1989a). Uniform asymptotic smoothing of Stokes's discontinuities. *Proc. R. Soc. Lond. A* **422**, pp. 7–21.

Berry, M.V. (1989b). Stokes phenomenon: smoothing a victorian discontinuity. *Pub. Math. de l'Institut des Hautes Études Scientifiques* **68**, pp. 211–221.

Berry, M.V. (1989c). Quantum Chaology, Not Quantum Chaos. *Physica Scripta* **40**, pp. 335–336.

Berry, M.V. and Howls C.J. (1991). Hyperasymptotics for integrals with saddles. *Proc. R. Soc. Lond. A* **434**, pp. 657–675.

Berry, M.V. (1992a). Asymptotics, superasymptotics, hyperasymptotics. in *Asymptotics beyond all orders, H. Segur and S. Tanveer Eds.* (Plenum, New York), pp. 1–14.

Berry, M V, (1992b). True Quantum Chaos? An Instructive Example. in New Trends in Nuclear Collective Dynamics, eds: Y Abe, H Horiuchi and K Matsuyanagi (Springer proceedings in Physics 58), pp 183-186.

Bertoncini, R., Kriman, A.M. and Ferry ,D.K. (1989). Airy-coordinate Green's function technique for high-field transport in semiconductors, *Phys. Rev. B* **40**, pp. 3371–3374.

Bertoncini, R., Kriman, A.M. and Ferry ,D.K. (1990). Airy-coordinate technique for nonequilibrium Green's function approach to high-field quantum transport, *Phys. Rev. B* **41**, pp. 1390–1400.

Besieris, I.M., Shaarawi, A.M. and Ziolkowski R.W. (1994). Nondispersive accelerating wave packets, *Am. J. Phys.* **62**, pp. 519–521.

Bethe, H.A. and Salpeter, E.E. (1957). Quantum mechanics of one and two electron atoms, Springer-Verlag, Berlin.

Bieniek, R.J. (1977). Uniform semiclassical methods of analyzing undulations in far-wing line spectra, *Phys. Rev. A* **15**, pp. 1513-1522.

Brault, P., Vallée, O. and Tran Minh, N. (1988). Non-perturbative uniform wavefunctions of coupled radial Schrödinger equations, *J. Phys. A* **21**, pp. L67–L73.

Brault, P., Vallée, O., Tran Minh, N. and Chapelle, J. (1988). Uniform semiclassical treatment of the radial coupling term in the adiabatic basis : application to the excitation transfer, *Phys. Rev. A* **37**, pp. 2318–2334.

Brillouin, L. (1916). Sur une méthode de calcul approché de certaines intégrales, dite méthode de col, *Ann. Sci. Éc. Norm. Sup.*, **33** pp. 17–69.

Brillouin, L. (1926). La mécanique ondulatoire de Schrödinger ; une méthode générale de résolution par approximations successives, *Compt. Rend. Acad. Sci. Paris* **183**, pp. 24–26.

Burnett, K. and Belsley, M. (1983). Uniform semiclassical off-shell wave functions and T-matrix elements, *Phys. Rev. A* **28**, pp. 3291–3299.

Chapman, A. (1992). George Biddel Airy, F.R.S. : a centenary commemoration, *Notes and Records R. Soc. London* **46**, pp. 103–110.

Chen, Z., Arce, P. and Locke, B.R. (1996). Convective–diffusive transport with a wall reaction in Couette flows, *The Chemical Engineering Journal* **61**, pp. 63–71.

Chen, Z. and Arce, P. (1997). An integral–spectral approach for convective–diffusive mass transfer with chemical reaction in Couette flow, *The Chemical Engineering Journal* **68**, pp. 11–27.

Chester, C., Friedman, B. and Ursell, F. (1957). An extension of the method of steepest descents, *Proc. Camb. Philos. Soc.* **53**, pp. 599–611.

Child, M.S. (1974). Molecular collision theory, Academic Press, London.

Child, M.S. (1975). A uniform approximation for one-dimensional matrix elements, *Mol. Phys.* **29**, pp. 1421–1429.

Clarkson, P.A. (2003). Painlevé equations—nonlinear special functions, *J. Comp. Appl. Math.* **153**, pp. 127–140.

Condon, E.U. (1928). Nuclear motion associated with electron transition in diatomic molecules, *Phys. Rev.* **32**, pp. 858–872.

Connor, J.N.L. (1979). Semiclassical theory of elastic scattering, in : Semiclassical methods in molecular scattering and spectroscopy, Proceedings of the NATO ASI (Cambridge, september 1979), M.S. Child Ed., Reidel, London.

Connor, J.N.L. (1980). Uniform semiclassical evaluation of Franck-Condon factors and inelastic atom-atom scattering amplitudes, *J. Chem. Phys.* **74**, pp. 1047–1052.

Connor, J.N.L. and Farrelly, D. (1980). Theory of cusped rainbows in elastic scattering : uniform semiclassical calculations using Pearcey's integral, *J. Chem. Phys.* **75**, pp. 2831–2846.

Connor, J.N.L., Curtis, P.R. and Farrelly, D. (1983). A differential equation method for the numerical evaluation of the Airy, Pearcey and swallowtail canonical integrals and their derivatives, *Mol. Phys.* **48**, pp. 1305–1030.

Copson, E.T. (1963). On the asymptotic expansion of Airy's integral, *Proc. Glasgow Math. Assoc.* **6**, pp. 113–115.

Copson, E.T. (1967). Asymptotic expansions, Cambridge University Press.

Crandal, R.E. (1996). On the quantum zeta function, *J. Phys. A:Math. Gen.* **29**, pp. 6795–6816.

Dando, P.A. and Monteiro, T.S. (1994). Quantum surface of section for the diamagnetic hydrogen atom: Husimi functions versus Wigner functions, *J. Phys. B* **27**, pp. 2681–2692.

Davies, B. (2002). Integral transform and their applications (third edition), Text in applied mathematics Vol 41, Springer Verlag, New York Berlin.

Davis, H.T. (1962). Introduction to nonlinear differential and integral equations, Dover.

Davison, S.G. and Glasser, M.L. (1982). Laplace transforms of Airy functions, *J. Phys. A* **15**, pp. L463–L465.

Drazin, P.G. and Reid, W.H. (1981). Hydrodynamic stability, Cambridge University Press.

Englert, B.G. and Schwinger, J. (1984). Statistical atom: Some quantum improvements, *Phys. Rev. A* **29**, pp. 2339–2352.

Erdélyi, A. (1956). Asymptotic expansions, Dover Publications, New York.

Erdélyi, A., Magnus, W., Oberhettinger, F., and Tricomi, F. (1981). Higher transcendental functions, Vol. III, Krieger Publishing Company, Malabar Florida.

Exton, H. (1985). The Laplace transform of the Macdonald function of argument $x^3/2$ and the Airy function $Ai(x)$, *IMA J. Appl. Math.* **34**, pp. 211–212.

Eu, B.C. (1984). Semiclassical theory of molecular scattering, Springer Verlag, Berlin.

Fabijonas, B.R. and Olver, F.W.J. (1999). On the reversion of an asymptotic expansion and the zero of the Airy functions, *SIAM Review* **41**, pp. 762–773.

Faxèn, H. (1921). Expansion in series of the integral $\int_y^\infty \mathrm{e}^{-x(t\pm t^{-\mu})} t^\nu \mathrm{d}t$, *Ark. Mat. Astronom. Fys.* **15**, No. 13, pp. 1–57.

Flajolet, P. and Louchard, G. (2001). Analytic variation on the Airy distribution, *Algorithmica* **31**, pp. 361–377.

Feng, M. (2001). Complete solution of the Schrödinger equation for the time-dependent linear potential, *Phys. Rev. A* **64**, 034101.

Fonck, R.J. and Tracy, D.H. (1980). Use of semiclassical wavefunctions for calculation of radial integrals in the Coulomb approximation, *J. Phys. B* **13**, pp. L101–L104.

Ford, J. (1989). Quantum chaos, is there any?, Directions in chaos, Vol. 2, Hao Bai-Lin Ed., World Scientific, Singapore.

Ford, J. and Ilg, M. (1992). Eigenfunctions, eigenvalues, and time evolution of finite, bounded, undriven quantum systems are not chaotic, *Phys. Rev. A* **45**, pp. 6165–6173.

Ford, J. and Mantica, G. (1992). Does quantum mechanics obey the correspondence principle? Is it complete?, *Am. J. Phys.* **60**, pp. 1086–1098.

Fuchs, L.I. (1884). Math. Werke **2** p. 355 (quoted by Ince (1956)).

Fusaoka, H. (1989). Common Airy function type solutions of some nonlinear equations, *J. Phys. Soc. Japan* **58**, pp. 1120–1121.

Gea–Banacloche, J. (1999). A quantum bouncing ball, *Am. J. Phys.* **67**, pp. 776–782.

Gil, A., Segura, J. and Temme, N. (2003). On the zeros of the Scorer functions, *J. Approximation Theory* **120**, pp. 253–266.

Gislason, E.A. (1973). Series expansion for Franck-Condon factors I. Linear potential and the reflexion approximation, *J. Chem. Phys.* **58**, pp. 3702–3707.

Goodmanson, D.M. (2000). A recursion relation for matrix elements of the quantum bouncer. Comment on "A quantum bouncing ball," by Julio Gea–Banacloche, *Am. J. Phys.* **68**, pp. 866–868.

Gordon, R.G. (1968). Error bounds in equilibrium statistical mechanics, *J. Math. Phys.* **9**, pp. 655–663.

Gordon, R.G. (1969). New method for constructing wavefunctions for bound states and scattering, *J. Chem. Phys.* **51**, pp. 14–25.

Gordon, R.G. (1970). Constructing wavefunctions for nonlocal potentials, *J. Chem. Phys.* **52**, pp. 6211–6217.

Gordon, R.G. (1971). Quantum scattering using piecewise analytic solutions, *Meth. in Comp. Phys.* **10**, pp. 81–109.

Gradshteyn, I.S. and Ryzhik, I.M. (1965). Tables of integrals, series and products, Academic Press, New York.

Gramtcheff, T.V. (1981). An application of Airy functions to the Tricomi problem, *Math. Nachr.* **102**, pp. 169–181.

Green, G. (1837). On the motion of waves in a variable canal of small depth and width, *Trans. Camb. Phil. Soc.* **6**, pp. 457–462.

Greenberger, D.M. (1980). Comment on "Nonspreading wave packets", *Am. J. Phys.* **48**, pp. 256.

Grémaud, B., Delande, D. and Gay J.C. (1993). Origin of narrow resonances in the diamagnetic Rydberg spectrum, *Phys. Rev. Lett.* **70**, pp. 1615–1618.

Hayasi, N. (1971). Higher approximations for transonic flows, *Quart. Appl. Math.* **29**, pp. 291–302.

Heller, E.J. (1976). Wigner phase space method : analysis for semiclassical applications, *J. Chem. Phys.* **65**, pp. 1289–1298.

Heller, E.J. (1984). Bound-state eigenfunctions of classically chaotic hamiltonian systems : scars of periodic orbits, *Phys. Rev. Lett.* **53**, pp. 1515–1518.

Hochstadt, H. (1973). Les fonctions de la physique mathématique, Masson, Paris.
Holschneider, M. (1995). Wavelets : An analysis tool, Oxford Science Publication.
Hunt, P.M. (1981). A continuum basis of Airy functions matrix elements and a test calculation, *Mol. Phys.* **44**, pp. 653–663.
Ince, E.L. (1956). Ordinary differential equations Dover, New York.
Iwasaki, K., Kajiwara, K. and Nakamura, T. (2002). Generating function associated with rational solutions of Painlevé II equation, *J. Phys. A: Math. Gen.* **35**, pp. L207–L211.
Jablonski, A. (1945). General theory of pressure broadening of spectral lines, *Phys. Rev.* **68**, pp. 78–93.
Jeffreys, H. (1923). On certain approximate solutions of linear differential equations of the second order, *Proc. London Math. Soc.* **23**, pp. 428–436.
Jeffreys, H. (1928). The effect on Love waves of heterogeneity in the lower layer, *Monthly Not. Royal Astron. Soc., Geophys. Supp.* **2**, 101–111.
Jeffreys, H. (1942). Asymptotic solutions of linear differential equations, *Phil. Mag.* **33**, pp. 451–456.
Kajiwara, K. and Ohta, Y. (1996). Determinant structure of the rational solutions for the Painlevé II equation, *J. Math. Phys.* **37**, pp. 4693–4704.
Kim Tuan, Vu (1998). Airy integral transform and the Paley-Wiener theorem, *Transform methods and special functions*; Proceedings of Varna'96: Bulgarian Academy of Sciences, pp. 523–531.
Knoll, J. and Schaeffer, R. (1977). Complex paths and uniform approximations in a semi-classical description of direct reactions, *Phys. Rep.* **31C**, pp. 159–207.
Kramers, H.A. (1926). Wellenmechanik und halbzählige Quantisierung, *Z. Phys.* **39**, pp. 828–840.
Krüger, H. (1979). Uniform approximate Franck–Condon matrix elements for bound-continuum vibrational transitions, *Theo. Chim. Acta* **51**, pp. 311–322.
Krüger, H. (1980). Approximate Franck–Condon factors from piecewise Langer transformed vibrational wave functions, *Theo. Chim. Acta* **57**, pp. 145–161.
Krüger, H. (1981). Semiclassical bound-continuum Franck–Condon factors uniformly valid at 4 coinciding critical points : 2 crossing and 2 turning points, *Theo. Chim. Acta* **16**, pp. 97–116
Landau, L. and Lifchitz, E. (1964). Théorie des champs, Mir, Moscou.
Landau, L. and Lifchitz, E. (1966). Mécanique quantique, Mir, Moscou.
Landau, L. and Lifchitz, E. (1967). Théorie de l'élasticité, Mir, Moscou.
Landau, L. and Lifchitz, E. (1971). Mécanique des fluides, Mir, Moscou.
Langer, R.E. (1931). On the asymptotic solutions of ordinary differential equations, with an application to the Bessel functions of large order, *Trans. Am. Math. Soc.* **33**, pp. 23–64.
Langer, R.E. (1955a). On the asymptotic forms of the solutions of ordinary differential equations of the third order in a region containing a turning point, *Trans. Am. Math. Soc.* **80**, pp. 93–123.
Langer, R.E. (1955b). The solutions of the differential equation $v''' + \lambda^2 w' + 3\mu\lambda^2 v = 0$, *Duke Math.* **22**, pp. 525–541.

Laurenzi, B.J. (1993). Moment integrals of powers of Airy functions, *ZAMP* **16**, pp. 891–908.

Leach, P.G.L. (1983). Laplace transforms of Airy functions via their integral definitions, *J. Phys. A* **16**, pp. L451–L453.

Lee, Soo-Y. (1980). The inhomogeneous Airy functions, Gi(z) and Hi(z), *J. Chem. Phys.* **72**, pp. 332–336.

Letellier, C. and Vallée, O. (2003). Analytic results and feedback circuit analysis for simple chaotic flows, *J. Phys. A: Math. Gen.* **36**, pp. 11229–11245.

Lermé, J. (1990). Iterative methods to compute one- and two–dimensional Franck–Condon factors. Test of accuracy and application to study indirect molecular transitions, *Chemical Physics* **145**, pp. 67–88.

Levin, E. and Lubinsky, D.S. (2009). On the Airy reproducing kernel, sampling series and quadrature formula, *Integr. eq. oper. theory* **63**, pp. 427–438.

Liouville, J. (1837). Sur le développement des fonctions ou parties de fonctions en séries. . . , *J. Math. Pures Appl.* **2**, pp. 16–35.

Littlejohn, R.G. and Cargo, M. (2002). An Airy discrete variable representation basis, *J. Chem. Phys.* **117**, pp. 37–42.

Lukes, T. and Somaratna, K.T.S (1969). The exact propagator for an electron in a uniform electric field and its application to Stark effect calculations, *J. Phys. C* **2**, pp. 586–592.

McDonald, S.W. and Kaufman, A.N. (1979). Spectrum and eigenfunctions for a hamiltonian with stochastic trajectories, *Phys. Rev. Lett.* **42**, pp. 1189–1191.

Martens, C.C., Waterland, R.L. and Reinhardt, W.P. (1988). Classical, semiclassical and quantum mechanics of a globally chaotic system : integrability in the adiabatic approximation, *J. Chem. Phys.* **90**, pp. 2328–2337.

Maslov, V.P. (1972). Théorie des Perturbations et Méthodes Asymptotiques, Dunod, Paris; Russian ed. (1965) Moscow.

Maurone, P.A. and Phares, A.J. (1979). On the asymptotic behavior of the derivatives of Airy functions, *J. Math. Phys.* **20**, pp. 2191–2191.

Meredith, D.C. (1992). Semiclassical wavefunctions of nonintegrable systems and localization on periodic orbits, *J. Stat. Phys.* **68**, pp. 97–130.

Miller, J.C.P. (1946). The Airy Integral, British Assoc. Adv. Sci. Mathematical Tables, Part-Volume B, Cambridge University Press, London.

Miller, W.H. (1968). Uniform semiclassical approximations for elastic scattering and eigenvalue problems, *J. Chem. Phys.* **48**, pp. 464–467.

Miller, W.H. (1970). Theory of Penning ionization, *J. Chem. Phys.* **52**, pp. 3563–3572.

Miller, W.H. (1975). The classical S-matrix in molecular collisions, *Adv. Chem. Phys.* **30**, pp. 77–136.

Miller, S.C. and Good, R.H. (1953). A WKB-type approximation to the Schrödinger equation, *Phys. Rev.* **91**, pp. 174–179.

Moon, W. (1981). Airy function with complex arguments, *Comput. Phys. Commun.* **22**, pp. 411–417.

Morse, M.P. and Feshbach, H. (1953). Methods of theoretical physics, Vol. I & II, Mac Graw Hill, Tokyo.

Moyer, C.A. (1973). On the Green's function for a particle in a uniform electric field, *J. Phys. C* **6**, pp. 1461–1466.

Nassar, A.B., Bassalo J.M.F. and de Tarso S. Alencar, P. (1995). Dispersive Airy packets, *Am. J. Phys.* **63**, pp. 849–852.

Nayfeh, A. (1973). Perturbation methods, Wiley, New York.

Neher, M. (1996). Validated bounds of the zeros of Airy functions, *Z. Angew. Math. Mech.* **79**, pp. S813–S814.

Nicholson, J.W. (1909). On the relation of Airy's integral to the Bessel functions, *Phil. Mag.* **18**, pp. 6–17.

O'Connell R.F. (1983). The Wigner Distribution Function–50th birthday, *Foundations of Physics*, **13**, pp. 83–92.

Olver, F.W.J. (1954a). The asymptotic solution of linear differential equations of the second order for large values of a parameter, *Phil. Trans. R. Soc. London A* **247**, pp. 307–327.

Olver, F.W.J. (1954b). The asymptotic expansion of Bessel functions of large order, *Phil. Trans. R. Soc. London A* **247**, pp. 328–368.

Olver, F.W.J. (1974). Asymptotics and special functions, Academic Press, New York.

Olver, P.J. (1998). Applications of Lie groups to differential equations, Graduate texts in Mathematics # 107, Springer Verlag.

Ozorio de Almeida, A.M. and Hannay, J.H. (1982). Geometry of two dimensional tori in phase space : projections, sections and the Wigner function, *Ann. Phys.* **138**, pp. 115–154.

Panda, S. and Panda, B. (2001). Analytic methods for field induced tunneling in quantum wells with arbitrary potential profiles, *Pramana–Journal of Physics* **56**, pp. 809–822.

Pearcey, T. (1946). The structure of an electromagnetic field in the neighbourhood of a cusp of a caustic, *Philos. Mag.* **37**, pp. 311–317.

Pechukas, P. (1972). Semiclassical approximation of multidimensional bound states, *J. Chem. Phys.* **57**, pp. 5577–5594.

Percival, I.C. (1973). Regular and irregular spectra, *J. Phys. B* **6**, pp. L229–L232.

Pike, E.R. (1964a). On the related-equation method of asymptotic approximation (W.K.B. or A-A method) I. A proposed new existence theorem, *Quart. J. Mech. Appl. Math.* **17**, pp. 105–124.

Pike, E.R. (1964b). On the related-equation method of asymptotic approximation (W.K.B. or A-A method) II. Direct solutions of wave penetration problems, *Quart. J. Mech. Appl. Math.* **17**, pp. 124–136.

Pomphrey, N. (1974). Numerical identification of regular and irregular spectra, *J. Phys. B* **7**, pp. 1909–1915.

Poston, T. and Stewart, I. (1978). Catastrophe Theory and its Applications, Pitman, Boston (and references therein).

Rabenstein, A.L. (1958). Asymptotic solutions of $u'''' + \lambda^2 (zu'' + \alpha u' + \beta u)$ for large $|\lambda|$, *Arch. Ratio. Mech. Analysis* **1**, pp. 418–435.

Reid, W.H. (1979). An exact solution of the Orr–Sommerfeld equation for plane Couette flow, *Stud. Appl. Math.* **61**, pp. 83–91.

Reid, W.H. (1995). Integral representations for products of Airy functions, *ZAMP* **46**, pp. 159–170.

Reid, W.H. (1997a). Integral representations for products of Airy functions Part 2. Cubic Products, *ZAMP* **48**, pp. 646–655.

Reid, W.H. (1997b). Integral representations for products of Airy functions Part 3. Quartic Products, *ZAMP* **48**, pp. 656–664.

Russell, J.S. (1844). Report on waves, Reports of the 14th meeting of the British Association for the Advancement of Science, London, pp. 311–390.

Salmassi, M. (1999). Inequalities satisfied by the Airy functions, *J. Math. Anal. Appl.* **240**, pp. 574–582.

Sando, K.M. and Wormhoudt, J.C. (1973). Semiclassical shape of satellite bands, *Phys. Rev. A* **7**, pp. 1889–1898.

Schulten, Z., Anderson, D.G.M. and Gordon, R.G. (1979). An algorithm for the evaluation of the complex Airy functions, *J. Comput. Phys.* **31**, pp. 60–75.

Scorer, R.S. (1950). Numerical evaluation of integrals of the form $I = \int_{x_1}^{x_2} f(x)\mathrm{e}^{\mathrm{i}\Phi(x)}\mathrm{d}x$ and the tabulation of the function $Gi(z) = (1/\pi)\int_0^\infty \sin\left(uz + u^3/3\right)\mathrm{d}x$, *Quart. J. Mech. Appl. Math.* **3**, pp. 107–112.

Soares, M., Vallée, O. and de Izarra, C. (1999). A study of the analytical and local semiclassical Wigner distribution, *Lecture Notes in Physics* – Vol. **518**, Dynamical systems, Plasmas and Gravitation, Springer Verlag, pp. 361–370.

Stokes, G.G. (1851). On the numerical calculation of a class of definite integrals and infinite series, *Trans. Camb. Phil. Soc.* **9**, pp. 166–187.

Stokes, G.G. (1858). On the discontinuity of arbitrary constants which appear in divergent developments, *Trans. Camb. Phil. Soc.* **10**, pp. 106–128.

Szudy, J. and Baylis, W.E. (1975). Unified Franck-Condon treatment of pressure broadening of spectral lines, *J. Quant. Spectros. Radiat. Transfer.* **15**, pp. 641–668.

Tabor, M. (1989). Chaos and integrability in non-linear dynamics, Wiley, New York.

Tellinghuisen, J. (1985). The Franck-Condon principle in bound-free transitions, *Adv. Chem. Phys.*, **LX**, pp. 299–369.

Temme, N.M. and Varlamov, V. (2009). Asymptotic expansions for Riesz fractional derivatives of Airy functions and applications, *J. Comp. Appl. Math.*, **232**, pp. 201–215.

Thom, R. (1975). Structural Stability and Morphogenesis. Benjamin–Addison Wesley, New York.

Titchmarsh, E.C. (1962). Eigenfunction expansions, Clarendon Press, Oxford.

Tomsovic, S. and Heller, E.J. (1993). Semiclassical construction of chaotic eigenstates, *Phys. Rev. Lett.* **70**, pp. 1405–1408.

Torres–Vega, Go., Zuniga–Secundo, A., and Morales–Guzman, J.D. (1996). Special functions and quantum mechanics in phase space: Airy functions, *Phys. Rev. A* **53**, pp. 3792–3797.

Tracy, C.A. and Widom, H. (1994). Level-spacing distributions and the Airy kernel, *Commun. Math. Phys.* **15**, pp. 151–174.

Trinkaus, H. and Drepper, F. (1977). On the analysis of diffraction catastrophes, *J. Phys. A* **10**, pp. L11–L16.

Turnbull, H.W. (1960). The theory of determinants, matrices and invariants, Dover Publications, New York.

Vallée, O. (1982). Uniform semiclassical evaluation of Franck-Condon factors in study of atom-diatom reactive collisions, Report #1567, Mechanical and Aerospace Engineering, Princeton, (unpublished).

Vallée, O., Soares, M. and de Izarra, C. (1997). An integral representation for the product of Airy functions, *ZAMP* **48**, pp. 156–160.

Vallée, O. (1999). On the linear third order differential equation, *Lecture Notes in Physics* – Vol. **518**, Dynamical systems, Plasmas and Gravitation, Springer Verlag, pp. 340–347.

Vallée, O. (2000). Comment on "A quantum bouncing ball," by Julio Gea–Banacloche, *Am. J. Phys.* **68**, pp. 672–673.

Vallée, O. (2002). Some integrals involving Airy functions and Volterra μ–functions, *Integral Transforms and Special functions*, **13**, pp. 403–408.

Varlamov, V. (2007). Semi-integer derivatives of the Airy function and related properties of the Korteweg–de Vries type equations, *Z. angew.. Math. Phys.*, **59**, pp. 381–399.

Varlamov, V. (2008a). Fractional derivatives of products of Airy functions, *J. Math. Anal. Appl.*, **337**, pp. 667–685.

Varlamov, V. (2008b). Differential and integral relations involving fractional derivatives of Airy functions and applications, *J. Math. Anal. Appl.*, **348**, pp. 101–115.

Varlamov, V. (2008c). Integral representations for products of Airy functions and their fractional derivatives, *Contemp. Math.*, **471**, pp. 203–218.

Voros, A. (1976). Semi-classical approximations, *Ann. Inst. Poincaré*, **24** pp. 31–90.

Voros, A. (1999). Airy function–exact WKB results for potentials of odd degree, *J. Phys. A: Math. Gen.* **32**, pp. 1301–1311.

Vrahatis, M., Ragos, O., Zafiropoulos, F.A., and Grapsa, T.N. (1996). Locating and computing zeros of Airy functions, *Z. Angew. Math. Mech.* **76**, pp. 419–422.

Watson, G. N. (1966). A treatise on the theory of Bessel functions, Cambridge University Press, London.

Wentzel, G. (1926). Eine Verallgemeinerung der Quantenbedingung für die Zwecke der Wellenmechanik, *Z. Phys.* **38**, pp. 518–529.

Widder, D.V. (1979). The Airy transform, *Am. Math. Month.* **86**, pp. 271–277.

Wigner, E.P. (1932). On the quantum correction for thermodynamic equilibrium, *Phys. Rev.* **40**, pp. 749–759.

Wille, L.T. (1986). Laplace transform of a class of *G* functions, *J. Phys. A* **19**, pp. L313–L315.

Wille, L.T. and Vennik, J. (1985). Evaluation of an integral involving Airy functions, *J. Phys. A* **18**, pp. 2857–2858.

Index

addition theorem, 82
adjoint equation, 118, 124
Airy averaging, 94
Airy equation, 5
Airy functions (generalisation of), 111–125
Airy kernel, 91, 145
Airy polynomials, 79, 81–83, 122, 125, 145
Airy transform, 73–94, 124, 182
Airy zeta function, 17–20
Albright (method of), 40–42
analytic continuation, 20
Appell set, 83
area distribution, 20
ascending series, 11–12, 28, 38
asymptotic expansion, 12–14, 23, 25, 29, 38–39
autocorrelation function, 80

Bessel functions, 20, 22, 27, 95, 112, 125, 158
binomial relation, 82

canonical form, 54, 62, 63, 67, 118
canonisation of cubics, 62–63
catastrophe, 114–117
Cauchy principal value, 56, 69, 79
causal relations, 69–70
caustic, 3, 114–117, 127–130
confluent hypergeometric function, 95
converging factor, 79
Couette flow, 134
Coulomb potential, 159–162

de Broglie wavelength, 156
Dirac delta function, 57, 64, 73
discriminant of cubics, 63

Ehrenfest theorem, 182
eigenfunction, 60, 74
Euler equation, 133

Faxén integral, 27
Fermi golden rule, 162
Fourier expansion, 76, 98
Fourier transform, 74, 80, 96, 131, 141, 183
Fourier–Airy series, 98
fractional derivative, 72–73
Franck–Condon factors, 162–170
Fredholm equation, 52, 90

Gauss integration method, 185–190
Gaussian function, 76
generating function, 83
golden mean, 65
Green's function, 70–71, 74

Hankel functions, 22
heat equation, 49, 137–139
Heisenberg, 171
Hermite polynomials, 83, 92
Hessian, 62

Hilbert transform, 69
Huygens principle, 127
hypergeometric function, 111

ill-posed problem, 80, 133
incomplete gamma function, 95
inequalities, 20
integral representation, 9–11, 27, 30
integrals, 50–67

Jacobian, 63
JWKB approximation, 104, 107, 163

Korteweg–de Vries equation, 60, 139–142

Laguerre polynomials, 159
Laplace (method of), 5, 120, 124
Laplace transform, 94
Legendre transformation, 131

Maar wavelets, 33
Meijer function, 95
Mellin transform, 20, 59, 62, 95
Mexican hat, 33
microcanonical distribution, 173–177
moments of Airy function, 52

Navier equation, 132

Orr–Sommerfeld equation, 124, 132–135
orthogonality condition, 74
orthonormal basis, 98
oscillating integrals, 101–104, 114–117

Painlevé equations, 143–145
parabolic cylinder function, 108
phase space, 171–173, 175, 178, 179, 182
Plancherel–Parseval, 75, 79, 85
Pochhammer symbol, 12, 24
Poiseuille flow, 124, 134
Poisson summation formula, 167, 187
Prandtl equation, 133
primitives, 37–49

quasi-static profile, 165

random phase approximation, 165
random walks, 19
recurrence relation, 60, 122, 123
Reynolds number, 124, 133, 135
Riccati equation, 34, 143, 188

scaling parameter, 75
Schrödinger equation, 74, 95, 147–162
Schwarzian derivative, 23, 105, 108, 119, 154
Scorer functions, 25, 29, 109, 123, 187
semiclassical limit, 154, 155, 157, 172, 175
semigroup, 73
similarity solution, 60, 61, 141
soliton, 139–142
stationary phase method, 101, 163, 164
steepest descent method, 12, 101
step function, 78
Stokes phenomenon, 14
syzygy, 63

Tchaplyguine equation, 130
third order differential equation, 30, 118
Thomas–Fermi, 94
transition point, 119
Tricomi equation, 130–132
turning point, 106, 107, 135, 155–159, 162, 163, 166, 173–178, 181

uniform approximation, 101–109, 154–162, 166

Volterra μ-function, 59–62

wavelets, 32–34
Weber functions, 22
Weierstrass infinite product, 17
Wronskian, 7, 30, 45, 118, 186

zeros of Airy functions, 16–20, 98